A Concise Human Biology and Hygiene

by

P. M. MINETT, B.SC.

MILLS & BOON LIMITED

First published in Great Britain 1973
by Mills & Boon Limited, 15–16 Brooks Mews,
London W1Y 1LF.

© P. M. Minett 1973

New material in 1979 edition, © P. M. Minett 1979
Reprinted (revised) 1975
Reprinted 1977
Reprinted (revised and extended) 1979
Reprinted 1980

ISBN 0 263 06412 3

Printed in Great Britain by
Thomson Litho Ltd, East Kilbride
and bound by Hunter & Foulis Ltd, Edinburgh

CONTENTS

ACKNOWLEDGEMENTS

For the 1979 edition of this book, I have to thank copyright holders for permission to reproduce:
Fig. 15.3, graphs from J. M. Tanner, *Growth at Adolescence,* 2nd edition, Blackwell Scientific Publications, 1962 (data from H. E. Jones, *Motor Performance and Growth,* University of California Press, 1949).
Fig. 15.5, diagrams from M. H. Day, *Fossil Man,* Hamlyn, 1969.
Fig. 15.6, graph from Samson Wright, *Applied Physiology,* 12th edition, Oxford University Press, 1971.
Several publications by the Health Education Council have been consulted with advantage.
My thanks are also due to the many specialists and colleagues whom I have consulted on various aspects of this work and whose friendly help and criticism have been invaluable.

P.M.M.
1979

PREFACE

This textbook has been written primarily for students preparing for the General Certificate of Education O Level examinations in Human Biology and Hygiene, or other examinations of a similar standard, and has been based on the syllabuses of the various Examining Boards. It provides more detail than is required for the Certificate of Secondary Education examinations, but will prove a useful reference source for those working for this examination. It also provides the background knowledge for the study of such subjects as nursing, public health, medicine, etc.

The book is intended as a textbook which presents the subject in a succinct, straightforward and factual manner, and which emphasises the important aspects and the fundamental framework of the subject without obscuring it with elaboration or specialist detail.

In the main a system of headings, sub-headings and notes has been used since the author feels that this will help the student to see the various aspects of the subject as a whole and to assimilate and memorise the details.

The book will prove particularly useful to those wishing to master the essentials of the subject in order to take the examination after a relatively short period of study, also as a consolidation or revision course for those who have made a more leisurely study over a period of several years. Those starting a sixth form course should find the book a good starting point for general study, discussion, research and experiment.

Preparation of a book of this type requires consultation with, and advice from, specialists and experts in many fields. The author would like to thank the many people who have so willingly and freely given of their time and knowledge. In particular, the author wishes to express gratitude to Dr R G Newberry, Medical Officer of Health, Great Yarmouth, and to Dr D H Perkins, also of Great Yarmouth.

P M M
July 1973

Chapter One

LIVING ORGANISMS

Human Biology is the study of man as a living organism. This involves the study of the structure and functioning of the human body, together with the nature of the environment and man's reaction to it, and his dependence on other living organisms for survival.

Living organisms are generally divided into two groups—plants and animals. It is usual to include bacteria and fungi in the plant kingdom although they do not possess all the characteristics typical of plants. Viruses appear to be intermediate between living organisms and non-living matter.

Man is a living organism and therefore possesses all the characteristics which distinguish living organisms from inanimate objects.

CHARACTERISTICS OF LIVING ORGANISMS

1. Respiration The process by which the body produces energy in order to carry out all the living processes (metabolic processes).* It usually involves the absorption of oxygen and the excretion of carbon dioxide.

2. Nutrition The process by which food is obtained so that it can be used either to provide energy or for growth and repair of the body.

3. Excretion The process by which the body gets rid of the waste products of metabolism.*

4. Growth The ability to increase in size.

5. Movement The ability to move. Animals can usually move from one place to another. Although some unicellular (one celled) plants can move about, most plants remain in one place. The movement shown by these plants include growth and response to light, water and other stimuli.

6. Sensitivity The ability to detect changes in the environment.

7. Reproduction The process by which new individuals are produced.

STRUCTURE OF THE CELL

Living organisms are composed of one or more cells. Each cell is a unit of life and contains living matter called **protoplasm.** Protoplasm is transparent, granular and jelly-like, either fluid or semi-solid, and it is in a constant state of movement. In a typical cell (Fig. 1.1) the protoplasm consists of **cytoplasm** and **nucleus.** The outer layer of cytoplasm forms the **cell membrane,** also called the

---

*Metabolism or **metabolic processes** are the terms applied to the complicated chemical changes that occur in living cells. The **metabolic rate** is the speed at which they occur.

1

plasma membrane, and it is **semi-permeable,** that is, it selectively allows some substances to pass through it but prevents the passage of others.

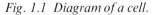
Fig. 1.1 Diagram of a cell.

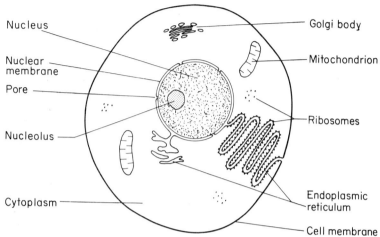

Cytoplasm is complex and contains various structures (**organelles**) each with its own part to play in the organisation of the cell. These include:

Mitochondria—rod-shaped bodies concerned with the energy supply of the cell. Here sugar is broken down (oxidised) and the energy which is released is stored in ATP (adenosine triphosphate) until it is required for the other activities of the cell (p. 37).

Endoplasmic reticulum—a complicated network of canals and vesicles (pockets) within the cytoplasm. The canals may link the exterior of the cell with the nucleus and be a passage for the transport of materials.

Ribosomes—granules which occur around the endoplasmic reticulum or free in the cytoplasm. They contain RNA and their function is to manufacture proteins. The cell requires a large number of different proteins to build up protoplasm and to make enzymes.

Golgi body—also called **golgi apparatus,** is always present but its function is not fully understood.

The **nucleus** is a densely packed, rounded structure embedded in the cytoplasm. It is surrounded by the **nuclear membrane** which contains small pores and it contains one or more spherical bodies called **nucleoli.** The nucleus contains the hereditary material (p. 91) and controls the activities of the cell. The chromosomes within the nucleus can only be seen when the cell is in the process of dividing. They possess the very special chemical called DNA (**deoxyribonucleic acid**). DNA molecules have the power to

1. Duplicate themselves during cell division so that the hereditary material can be passed on to daughter cells.

2. Make molecules of RNA (**ribonucleic acid**). The RNA molecules act as messengers and pass through the nuclear membrane into the cytoplasm with instructions for the chemical activities of the cell.

Plant cells differ from animal cells

(i) The cytoplasm secretes a **cellulose cell wall** around the cell.

(ii) In the centre of the cell a space called the **vacuole** is filled with **cell sap.**

(iii) Many cells possess **chloroplasts** containing the green pigment **chlorophyll,** which gives green plants their colour.

CELL DIVISION

New cells are formed from existing cells by cell division. This enables the organism to grow and develop and to replace damaged tissue. During the usual form of cell division the nucleus of the cell divides into two by a process called mitosis, this is followed by division of the cytoplasm.

Mitosis

The division of the nucleus which results in the daughter cells each having a nucleus with the same number of chromosomes as the parent cell is called mitosis. This type of division ensures that all the cells of an organism have the same hereditary material and will behave in a characteristic way for that organism e.g. all the cells in the body of a man have the hereditary material of a man and no other organism.

Mitosis is an orderly process and the following stages can be recognised. (N.B. for clarity only two chromosomes are shown in the diagrams.)

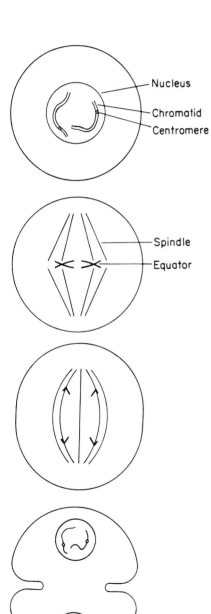

1. In a cell that is about to divide the chromosomes become visible in the nucleus. Each consists of two parallel threads called **chromatids** joined by a **centromere.**

2. The nuclear membrane disappears. Threads of protein form a **spindle** and the chromosomes shorten and thicken and line up along the **equator.**

3. The two halves of each chromosome separate and go to opposite ends of the cell.

4. A nuclear membrane forms around each group of chromosomes and the cytoplasm divides. The two daughter cells thus formed each contain the same number of chromosomes as the original cell. The chromosomes lengthen and disappear and the new cells grow.

DIFFERENCES BETWEEN PLANTS AND ANIMALS

The fundamental difference between plants and animals is in their method of nutrition. Green plants make their own food by the process of photosynthesis, whereas animals obtain their food by eating it, then digesting and absorbing it.

Photosynthesis

Photosynthesis is the process by which green plants use energy from sunlight to manufacture sugar from carbon dioxide and water.

It can be summed up by the equation

$$6CO_2 + 6H_2O + \text{sunlight} \xrightarrow{\text{chlorophyll}} C_6H_{12}O_6 + 6O_2$$

carbon   water   glucose containing oxygen
dioxide   stored energy

In order that photosynthesis can take place, the cells must contain the green pigment chlorophyll, be exposed to sunlight, and have a plentiful supply of carbon dioxide and water. Oxygen is given off as a waste product. The carbohydrates which are formed are simple sugars such as glucose and they contain stored energy from the sun. This energy is released when the sugar is used in respiration. However, the plant may use the sugar to form complex carbohydrates such as starch or cellulose, or convert it into fat or oil (e.g. olive oil), or it might be combined with nitrogen and other elements obtained from the soil to form protein.

Summary

1. Plants manufacture food by photosynthesis. Animals obtain food by eating.

2. Plants contain the green pigment chlorophyll. Animals do not.

3. Plants have cellulose cell walls. Animals do not.

4. Plants are held in one place by roots so that water and mineral salts can be absorbed from the soil. Animals move from place to place in search of food.

5. Plants do not have a nervous system. Animals need one so that they can be aware of changes in the environment as they move around.

6. Plants give off oxygen during the day (photosynthesis) and carbon dioxide at night (respiration). Animals excrete carbon dioxide continuously, but never oxygen.

N.B. Although plants respire continuously during the day, the carbon dioxide thus produced is used for photosynthesis.

7. Growth in length of the plant is at the tips of the branches of roots and stem. Growth in length of an animal is because all parts of the body enlarge.

INTERDEPENDENCE OF PLANTS AND ANIMALS

Animals need plants

1. For food to provide energy. Plants have the ability to store energy in carbohydrate during photosynthesis and animals need this energy for all their metabolic processes.

2. For food to supply materials for growth and repair of body tissues. Plants have the ability to convert carbohydrate into proteins using nitrogen and other elements they have absorbed from the soil. Animals need proteins to build up muscle and other tissues and they get these by eating plant protein, or by eating animals which have fed on plants e.g. man eats beef from cattle fed on grass.

3. For oxygen. Plants give off oxygen into the air as a waste product of photosynthesis.

Plants need animals

1. For carbon dioxide. Animals continuously give off carbon dioxide, which the plants need in order to photosynthesise.

2. For a supply of nitrates from which to obtain nitrogen to build up protein. Nitrates are released into the soil from decaying animal matter, either excreta or dead tissue.

Food Chains

A food chain illustrates that all food ultimately comes from plants. E.g. (i) man eats beef, milk, butter and cheese, but as cattle feed on grass therefore the beef, milk, butter and cheese must have originated as grass.

(ii) man eats large fish which eat smaller fish which eat plankton, and plankton contains minute plants.

THE NITROGEN CYCLE

The continuous circulation of nitrogen in different chemical compounds as the result of activity by living organisms (Fig. 1.3).

Although about 79% of air is nitrogen, it cannot be used in this form by either plants or animals. Plants obtain their nitrogen in the form of nitrates from the soil. Animals obtain their nitrogen by eating plant or animal protein. Bacteria play a vital part in breaking down dead organic matter (humus) to release the nitrates from protein and urea.

Fig. 1.3 The nitrogen cycle. The chemical compounds containing nitrogen are shown in capital letters.

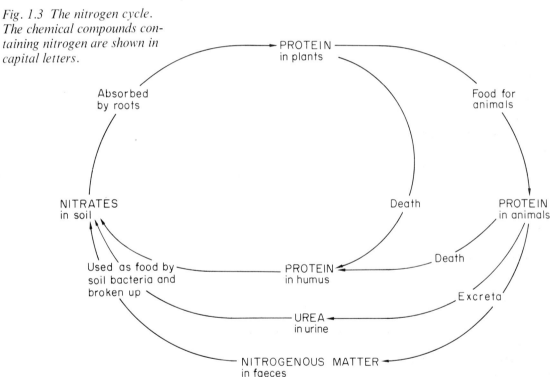

THE CARBON CYCLE

The continuous circulation of carbon in different chemical compounds as the result of activity by living organisms (Fig. 1.4).

Animals obtain their carbon in food—carbohydrates, fats and protein all contain carbon. Plants obtain their carbon in the form of carbon dioxide from the air in order to build up carbohydrates which can then either be used for energy or be converted into other substances to build the plant body. The carbon dioxide content of the air is kept at a more or less constant level as both plants and animals continuously excrete it as a waste product of respiration. Bacteria also release carbon dioxide as they break down the humus.

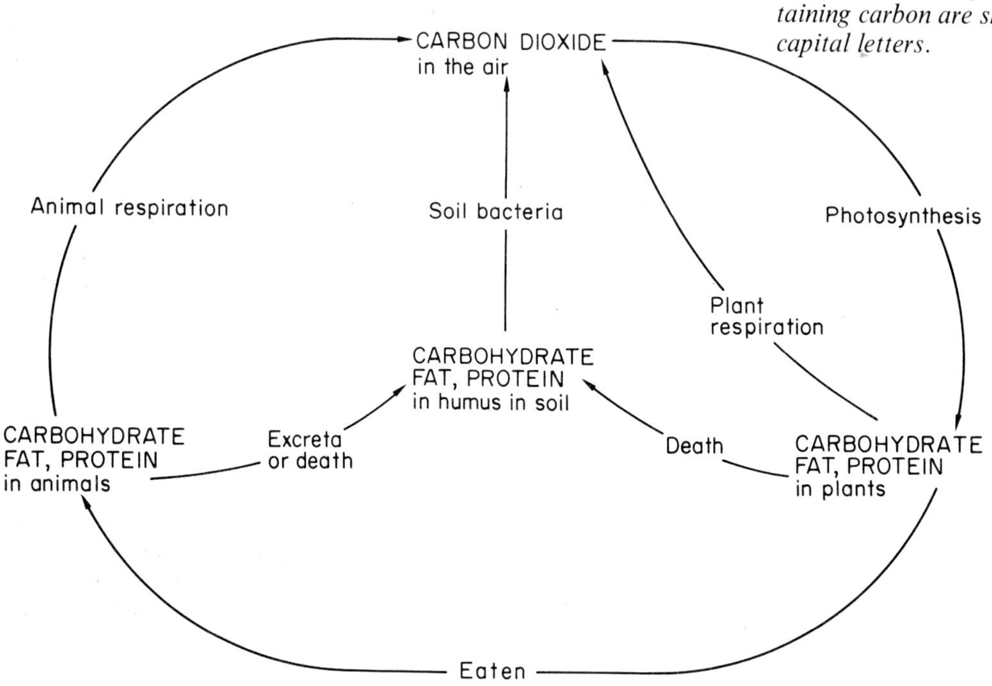

Fig. 1.4 The carbon cycle. The chemical compounds containing carbon are shown in capital letters.

CARBON DIOXIDE
in the air

Animal respiration   Soil bacteria   Photosynthesis

Plant respiration

CARBOHYDRATE
FAT, PROTEIN
in humus in soil

CARBOHYDRATE   Excreta   Death   CARBOHYDRATE
FAT, PROTEIN   or death   FAT, PROTEIN
in animals   in plants

Eaten

BACTERIA

Bacteria are micro-organisms that are found almost everywhere— in the air, soil, water, food, on or in plants and animals, and in humus. **Humus** is the dead organic matter in soil such as dead insects and worms, faeces, leaves and old roots. The structure of bacteria is described on p. 112 and an experiment to detect the presence of bacteria in the environment is described on p. 117.

There are vast numbers of different types of bacteria and each species has its own conditions in which it can live and grow. Should the conditions become unfavourable these organisms have the power to remain inactive until the environment is again suitable for growth. Some species produce spores that are capable of withstanding extremely unfavourable conditions which would kill non-sporing bacteria.

Moisture

Moisture is essential and most bacteria are capable of living in water. They are all inactivated or killed by very dry conditions.

Temperature

Most bacteria grow best at moderate temperatures but some prefer either high or low ones. Temperatures above 60 °C (140 °F)

usually kill bacteria, but low temperatures usually only inhibit their reproduction. Some spores, however, can survive in boiling water for several hours, while others can survive temperatures below freezing for many years.

Sunlight

Sunlight often slows down bacterial growth and ultra-violet rays can kill them. Greatest bacterial activity takes place in the dark, such as in soil or rubbish dumps.

Food

Different bacteria require different foods, some being very specific, others are able to feed on a wide range of material. As one species of bacterium finishes with a food, another will feed on the remains. In this way complex organic matter gradually becomes broken down into its inorganic parts by a succession of bacteria and other micro-organisms.

Oxygen

Aerobic Bacteria require oxygen obtained from air in order to respire like plants and animals.

Anaerobic Bacteria will not grow in the presence of oxygen. This accounts for the great amount of bacterial activity that takes place within rubbish dumps and similar places into which air cannot penetrate. These are the bacteria which produce foul-smelling gases.

The Importance of Bacteria to Man

Bacteria Harmful to Man

Mostly those which cause disease and they are dealt with in Chapter 13. Other harmful bacteria turn food bad.

Bacteria Beneficial to Man

1. Those which live in the large intestine and produce vitamin K.
2. Those which live on humus and by using it as food eventually break it down into inorganic particles. These provide food for plants which in turn provide food for man. These bacteria are an essential part of the nitrogen cycle (Fig. 1.3).
3. The bacteria which live on humus also help to prevent the accumulation of rubbish. These bacteria are therefore important in the treatment of sewage (p. 151) and of household refuse dumped in pits (p. 153).
4. Some bacteria are useful in food production, for example they turn

    milk into cheese
    milk into yoghurt
    cabbage into sauerkraut

5. Other bacteria are used in industrial processes, for example
    curing of tea leaves
    curing of tobacco leaves
    tanning of leather

FUNGI

Fungi are simple plants which do not contain chlorophyll. Therefore as they cannot photosynthesise they obtain their food by either being parasitic on other living organisms or by obtaining food from dead plant or animal matter. There are thousands of different fungi including yeasts, moulds, mushrooms, toadstools, puff-balls. The fungus body is made up of fine branching threads and reproduction is by spores which can easily be dispersed by air currents.

Importance of Fungi to Man

Fungi Harmful to Man

1. Those which are poisonous when eaten.
2. Those which cause skin disease, e.g. ringworm, athlete's foot, thrush.
3. Those which damage food crops, e.g. potato blight, 'smut' of wheat.
4. Those which turn food bad, e.g. mould on bread, jam and oranges.
5. Those which damage useful materials, e.g. dry-rot in wood, mould on leather.

Fungi Beneficial to Man

1. Those which can be eaten as food, e.g. mushrooms, blue mould in cheese.
2. Those which, together with bacteria, assist the decay of humus and prevent the accumulation of rubbish.
3. Yeasts, used in the baking of bread and the fermentation of alcohol.
4. *Penicillium* and other moulds from which antibiotics such as penicillin and streptomycin are obtained.

CELLS, TISSUES, ORGANS, SYSTEMS

Cells

A cell is a unit of life and in the smaller and simpler animals (e.g. *Amoeba*) and plants (e.g. *Spirogyra*) each cell is capable of carrying out all the activities of life. But in the more complex multicellular organisms, cells develop differently in order to specialise in a particular function. For example, in man cells vary considerably in size, shape and behaviour, but with each type of cell their structure is related to their function. Some examples are

   (i) muscle cells (Fig. 3.7) contain contractile material so that they can contract and cause movement.
  (ii) bone cells (Fig. 3.1) secrete a very hard substance to give the skeleton strength.
 (iii) cartilage cells (Fig. 3.3) secrete a tough, flexible substance so that certain parts of the skeleton are less brittle.
  (iv) nerve cells (Figs. 8.1 and 8.2) have long processes in order to convey messages quickly from one part of the body to another.
   (v) germinative cells of the epidermis remain undifferentiated and continually divide to form a protective layer on the outside of the body.

8

Tissues

A tissue is a group of cells of similar form and function. Tissues can be classified into four groups:

1. *Muscle tissue*
   (a) smooth
   (b) cardiac
   (c) skeletal
2. *Connective tissue*
   (a) bone
   (b) cartilage
   (c) blood
   (d) ordinary connective tissue*
3. *Nerve tissue*
   (a) neurones
   (b) sheath cells
4. *Epithelial tissues*—these cover the surfaces of the body, both external and internal. The functions are protection and secretion.
   (a) epidermis
   (b) mucous membrane**

Organs

An organ is comprised of more than one kind of tissue and it is a part of the body with a special function or functions. Examples are

Kidney for excretion and water-regulation
Lung for breathing
Heart for pumping blood.

Systems

A system is a group of organs all working in harmony so that the body can carry out a particular function or functions.

E.g. Digestive system. In order to carry out its functions of eating, digesting and absorbing food, it involves the following organs— teeth, mouth, oesophagus, stomach, intestines, liver, pancreas.

* Ordinary connective tissue (usually just called **connective tissue**) is smooth and moist and binds together the various parts in the body. It contains white fibres to give the tissue strength, and yellow fibres to give it elasticity so that movement can take place.
** Mucous membrane contains glands that secrete mucus to keep the surface moist. Many cells of the mucous membrane of the respiratory tract and Fallopian tubes have cilia, that is, minute hair-like outgrowths. The cilia beat rhythmically to push the mucus along.

Chapter Two

EVOLUTION AND MAN

EVOLUTION

Evolution is the name given to the way in which simple forms of life, by slow changes, have given rise to higher, more complex forms of life.

Geologists have estimated that the earth is approximately five billion years old and that life has been present during the last two billion years. It is probable that the first forms of life were simple and unspecialised, but even so, they must have contained the basic material of life called protoplasm. From these early forms developed the two great groups of living organisms—plants and animals.

The earliest forms of animal life to develop lacked a backbone, they are called **invertebrates.** Later, animals with a backbone evolved and they are called the **vertebrates.** Fish were the first vertebrates to evolve, amphibians evolved from fish, and reptiles from amphibians. These are all cold-blooded. The reptile stock gave rise to the two warm-blooded groups—birds and mammals.

During the last fifty million years or so mammals have continued to evolve and different groups have emerged including marsupials, anteaters, rodents, bats, elephants and primates. The **primate** group includes tree-shrews, bush babies, monkeys, apes and man, and they all evolved from primitive primate stock. Man is more like the ape than any other type of animal, but differs from apes in that he stands upright, the legs are longer than the arms, he walks and runs on his hind legs leaving the hands free for other tasks, and he has a larger brain and a much greater ability to learn and to communicate.

The first men lived over one million years ago and, although called **ape men,** they were more like men than apes. They were gradually replaced by **true man** who could make tools, use fire and cook. These were succeeded by **modern man** who appeared about 40 000 years ago. His large brain and his ability to learn enabled modern man to make rapid progress. He learnt to alter his environment and by 9000 BC he was farming and building cities.

VARIETIES OF MAN

Human beings belong to the species *Homo sapiens* and all have similar organs and bodily functions, but within the species there are variations in appearance. One method used to sub-divide the species into smaller groups called races is based largely on skin colour and hair. There is a rough correlation between geographical areas and physical characteristics.

1. Caucasoid (Europe and West Asia)—light skin, straight or wavy hair.

2. Mongoloid (Eastern Asia)—yellowish skin, straight black hair.

3. Negroid (Africa, below the Sahara)—black skin, curly hair.

4. Asians (Southern and South-East Asia)—brown skin, straight black hair.

5. American Indian (America)—reddish skin, straight black hair.

EVIDENCE OF EVOLUTION

1. Fossils—the remains or traces of plants or animals which have been preserved in rock. These organisms were living at the time the rocks were being formed and most are now extinct. The study of fossils reveals the types of animals and plants which have previously existed and the sequence in which they evolved.

2. Comparative anatomy—the comparison of the detailed structure of animals and plants. If the structure of two different animals has a common plan, it suggests that they have a common ancestor.

3. Embryology Animals which look markedly different in the adult stage may have a similar embryo stage. For example, the human embryo has gill slits and a tail, and it resembles the embryos of most other vertebrates. This suggests that they are all descended from a common ancestor.

4. Plant and animal breeding Breeding experiments show that large numbers of variations can occur within a species, e.g. the many varieties of wheat, or cattle. In nature, it is likely that when variations occur, those which give the organism a greater chance of survival are likely to be passed on to future generations, while the unfavourable variations would die out. Darwin called this 'survival of the fittest'. In time, as the new varieties developed other variations to cope with changing environmental conditions, the resulting organisms may be completely different from the original. In this way new species evolve.

MAN AS A LIVING ORGANISM

Man is a living organism and therefore possesses those characteristics attributed to living organisms described on p. 1.

Living organisms are classified as plants or animals. Man is an animal as he possesses the characteristics of animals which are summarised on p. 4.

There are various groups into which animals are classified and man is regarded as a **mammal** because he possesses the characteristics of this group which are

1. The presence of hair.

2. The presence of sweat glands.

3. The young develop within the uterus and are supplied with food and oxygen by the mother through the placenta.

4. After birth the mother feeds the young on milk secreted by the mammary glands.

5. A diaphragm separates thorax from abdomen.

6. The heart is divided into four chambers.

CHARACTERISTICS OF MAN

Man differs from other mammals because

1. The brain is relatively very large and highly developed giving

a much greater degree of intelligence. Hence man is
  (a) inquisitive, inventive and learns easily from experience.
  (b) aware of his actions and able to think.
  (c) able to communicate with other people by speech and music, drawing, writing and reading.
  2. The posture is naturally upright with two legs only being used for walking.
  3. The arms have developed other skills. The hands are very efficient due to the fact that the thumbs can move to oppose the fingers. This allows the hand to grasp objects easily.
  4. Man makes tools.
  5. The body is covered by relatively little hair.

GENERAL STRUCTURE OF MAN

The body consists of head, neck, trunk, two arms and two legs.

Head

The brain occupies about half the space within the head and nearby are the majority of the sense organs—eyes, ears, nose and tongue. All are protected by the bony skull. The nose is also the opening for the respiratory system. The lower jaw aids speech and because it can move sideways it enables chewing.

Neck

This allows movement of the head and contains the pharynx and larynx. It is the region through which the spine and oesophagus pass.

Trunk

This is divided into two cavities by the **diaphragm** which is a sheet of muscular tissue used in breathing. The upper cavity is called the **thorax** and contains heart and lungs, the lower cavity is the **abdomen** which contains the main organs of digestion, excretion and the female organs of reproduction.

THE STUDY OF MAN

  Anatomy is the study of the structure of the body.
  Histology is the detailed study of the cells and tissues.
  Physiology is the study of how the body functions.
  Pathology is the study of disease.
  Hygiene is the study of the principles governing health.

Chapter Three

THE SKELETON AND MUSCLES

SKELETON

The skeleton is a strong internal framework of bone and cartilage.

FUNCTIONS

1. Provides an anchorage for skeletal muscles which allows the body to move freely.
2. Protects the delicate organs within the cranium and thorax.
3. Supports the other parts of the body.
4. The bone marrow produces red blood cells.

In the early stages of development of the embryo the skeleton is composed of cartilage, but as growth continues it is largely replaced by bone. Bone and cartilage are living tissues which secrete non-living substances to give them strength.

BONE

Bone is a hard tissue. Canals called **haversian canals** which contain blood vessels and nerve fibres penetrate the bone tissue, and the bone cells are arranged concentrically around them. The hard substance of bone, the **matrix,** is secreted by the bone cells and contains **collagen** fibres and calcium compounds. The bone cells interconnect with each other and the haversian canals by minute canals called **canaliculi** through which tissue fluid can pass. (Fig. 3.1.)

Fig. 3.1 A cross-section through bone tissue.

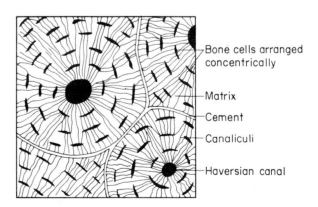

Bone cells arranged concentrically

Matrix

Cement

Canaliculi

Haversian canal

Bone is referred to as **compact** when it lacks spaces within the tissue. It is called **spongy bone** when there are interconnecting spaces filled with **bone marrow.** Compact bone surrounds spongy bone and a long bone will have a central cavity also filled with marrow (Fig. 3.2). The marrow is responsible for the production of all the red blood cells and some white blood cells.

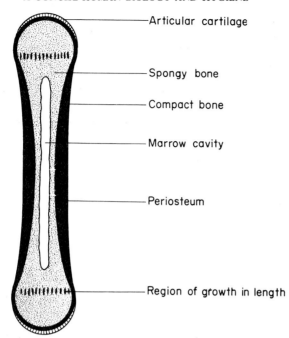

—Articular cartilage

—Spongy bone

—Compact bone

—Marrow cavity

—Periosteum

—Region of growth in length

Fig. 3.2 L.S. typical long bone.

As bone is a living tissue it is capable of growth and repair and because of this broken bones can be mended. Throughout life the calcium compounds in the hard matrix are continuously being removed and replaced.

The **periosteum** is the tough membrane which surrounds a bone and forms a strong union with tendons and ligaments. It also supplies blood vessels and nerves to the bone, and the cells immediately beneath it enable the bone to grow in thickness. Growth in length of a long bone takes place in special regions near the ends of the bone.

CARTILAGE

Cartilage is a tough but flexible tissue. It lacks blood vessels but receives nourishment from the surrounding tissues. Cartilage cells occur singly or in small groups in cavities in the matrix which they have secreted to give the tissue its strength and flexibility (Fig. 3.3).

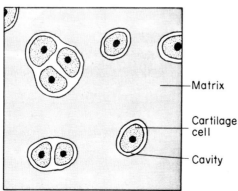

—Matrix

Cartilage cell

Cavity

Fig. 3.3 A cross-section through cartilage tissue.

There are three main types:

1. **Hyaline Cartilage** is semi-transparent and forms **articular cartilage** at the end of bones where they form movable joints, also the cartilage of the larynx and nose and the C-shaped cartilages of the trachea.

2. **White Fibro Cartilage** contains tough white fibres to give extra strength. This forms the intervertebral discs.

3. **Yellow Fibro Cartilage** contains yellow elastic fibres to allow it to stretch. This forms the lobe of the ear and the epiglottis.

Fig. 3.4 Joints.

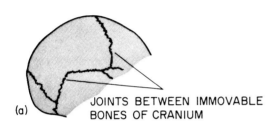

(a) JOINTS BETWEEN IMMOVABLE BONES OF CRANIUM

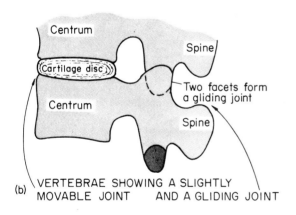

(b) VERTEBRAE SHOWING A SLIGHTLY MOVABLE JOINT AND A GLIDING JOINT

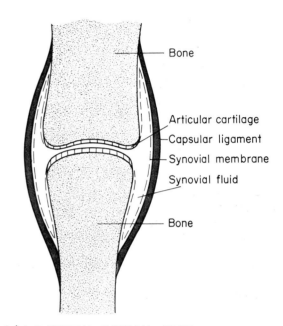

(c) L.S. TYPICAL SYNOVIAL JOINT

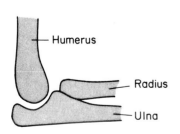

(d) HINGE JOINT OF ELBOW
(capsular ligament omitted)

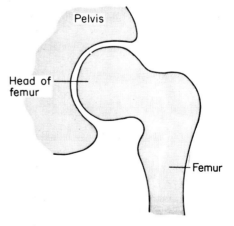

(e) BALL AND SOCKET JOINT OF HIP
(capsular ligament omitted)

LIGAMENTS

These are bands of strong, flexible fibrous tissue containing both tough and elastic fibres. Ligaments help to hold the bones in position at the joints. When a ligament completely encircles the joint it is called a **capsular ligament.**

JOINTS

A joint is where two bones meet. Joints are classified according to the amount and type of movement between the two bones.

1. **Immovable Joints** are those where there is no movement between the bones, e.g. (a) between bones of the cranium (Fig. 3.4.a); (b) between teeth and jaw bone.

2. **Slightly Movable Joints** are those when only very little movement can take place between two bones. This movement is possible because a pad of cartilage separates the two bones and this has a certain amount of flexibility. E.g. (a) between the centra of two vertebrae (Fig. 3.4.b); (b) between the two pubic bones.

3. **Freely Movable Joints** are also called **Synovial Joints** (Fig. 3.4.c). These allow considerable freedom of movement between the two bones because the ends of the bones are covered with cartilage to give a smoothness, and the whole joint is completely enclosed by the **capsular ligament.** This is lined with **synovial membrane** which secretes **synovial fluid** to lubricate the joint. As one bone articulates with the other, the cartilage at the ends of the bones is called **articular cartilage.** There are various types of synovial joints.

  (i) *Gliding Joint*   This allows one bone to glide over another; e.g. (a) between bones of wrist and ankle; (b) between facets of adjacent vertebrae (Fig. 3.4.b).

 (ii) *Hinge Joint*   This allows movement in one plane only; e.g. elbow joint (Fig. 3.4.d), knee joint.

(iii) *Ball and Socket Joint*   This allows movement in any plane as the end of one bone is rounded and fits into a socket of the other bone, e.g. shoulder joint, hip joint (Fig. 3.4.e).

(iv) *Pivot Joint*   This allows movement of one bone around another. The atlas pivots around the odontoid process of the axis to allow the head to be turned from side to side.

The **skeleton** consists of over 200 bones which make up the vertebral column, the ribs and sternum, skull, pectoral girdle, pelvic girdle, arms and legs (Fig. 3.5).

THE VERTEBRAL COLUMN

The vertebral column consists of a column of small bones, called **vertebrae.**

Functions

1. It supports the body in an upright position.
2. It encloses and protects the spinal cord.
3. It provides points of attachment for the powerful muscles of the back.
4. The joint between the atlas and skull allows the head to move backwards and forwards.

5. The joint between the atlas and axis allows the head to turn from side to side.

6. The joints between the other vertebrae allow a small amount of movement, and the combined movement of all the separate joints allows for considerable bending and rotation of the trunk.

Fig. 3.5 The skeleton.

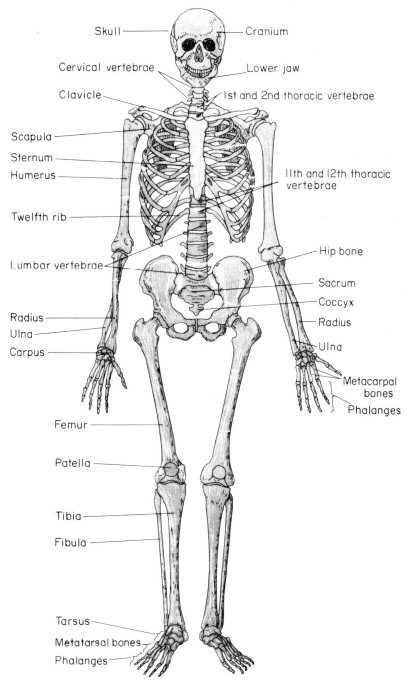

Skull — Cranium

Cervical vertebrae — Lower jaw

Clavicle — 1st and 2nd thoracic vertebrae

Scapula

Sternum

Humerus — 11th and 12th thoracic vertebrae

Twelfth rib

Hip bone

Lumbar vertebrae — Sacrum

Coccyx

Radius — Radius

Ulna — Ulna

Carpus — Metacarpal bones

Phalanges

Femur

Patella

Tibia

Fibula

Tarsus

Metatarsal bones

Phalanges

Vertebrae

A typical vertebra, of which a lumbar vertebra is a good example, has the following features:

A **centrum** which forms the main part of the bone.

A **neural canal** through which the spinal cord passes.

A **spine** which projects dorsally ⎫ for the attachment
Two **transverse processes** which ⎬ of muscles.
project laterally ⎭

Facets for articulation with other bones.

There are 33 vertebrae and in different parts of the column they are modified for different functions (Fig. 3.6).

7 Cervical Vertebrae form the neck. They are relatively small and can be recognised by the hole in each of the transverse processes through which passes the vertebral artery. Also the spine is forked. The **Atlas** is the first cervical vertebra and has large facets to articulate with the skull. It also has a large opening to accommodate the odontoid process of the axis. The **axis** is the second cervical vertebra.

12 Thoracic Vertebrae are larger and can be recognised by the two extra facets on each side which articulate with the ribs.

5 Lumbar Vertebrae are the largest and strongest to support the extra weight above, and for the attachment of the powerful muscles of this region.

The sacrum is formed from the fusion of five vertebrae and is roughly triangular in shape.

The coccyx is formed from the fusion of 4 vertebrae. It is very small, in fact it is the remnants of a tail.

THE RIBS AND STERNUM

The ribs and sternum form a cage around the thoracic cavity. There are 12 pairs of ribs which articulate with the 12 thoracic vertebrae. 10 pairs encircle the thorax to be joined by cartilage to the sternum (breast bone), the other two pairs are short and are attached only to the vertebral column, they are called **floating ribs.**

Functions

1. To protect the heart and lungs.
2. Their movement assists breathing.

THE SKULL

The skull consists of the cranium with cavities for eyes, ears and nose, a movable lower jaw, and teeth.

Functions

1. To protect the brain.
2. To protect the sense organs—eyes, nose, ears, and tongue.
3. To assist with talking and breathing and chewing.

THE PECTORAL GIRDLE

The pectoral girdle (shoulder girdle) consists of 2 shoulder blades and 2 collar bones.

Fig. 3.6 *The vertebral column.*

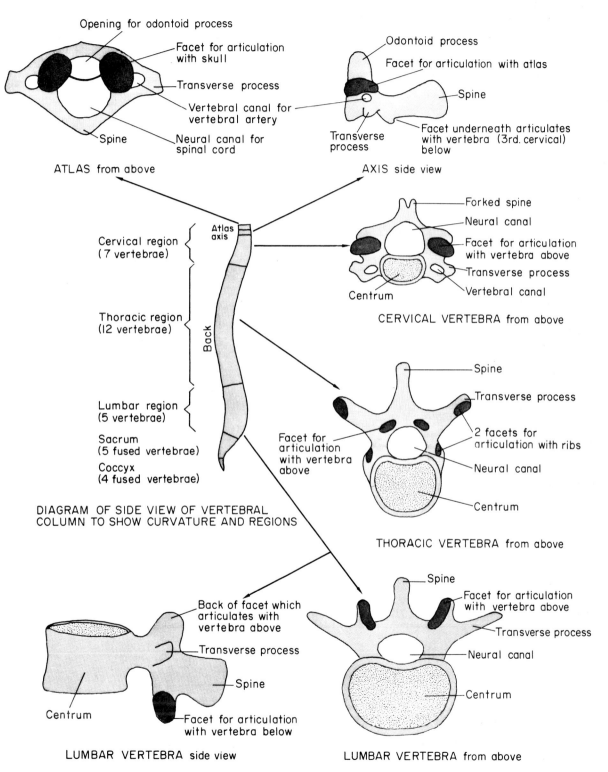

Opening for odontoid process

Facet for articulation with skull

Transverse process

Vertebral canal for vertebral artery

Spine

Neural canal for spinal cord

ATLAS from above

Odontoid process

Facet for articulation with atlas

Spine

Transverse process

Facet underneath articulates with vertebra (3rd. cervical) below

AXIS side view

Atlas axis

Cervical region (7 vertebrae)

Thoracic region (12 vertebrae)

Back

Lumbar region (5 vertebrae)

Sacrum (5 fused vertebrae)

Coccyx (4 fused vertebrae)

DIAGRAM OF SIDE VIEW OF VERTEBRAL COLUMN TO SHOW CURVATURE AND REGIONS

Forked spine

Neural canal

Facet for articulation with vertebra above

Transverse process

Vertebral canal

Centrum

CERVICAL VERTEBRA from above

Spine

Transverse process

2 facets for articulation with ribs

Neural canal

Facet for articulation with vertebra above

Centrum

THORACIC VERTEBRA from above

Back of facet which articulates with vertebra above

Transverse process

Spine

Centrum

Facet for articulation with vertebra below

LUMBAR VERTEBRA side view

Spine

Facet for articulation with vertebra above

Transverse process

Neural canal

Centrum

LUMBAR VERTEBRA from above

Functions

1. To form ball and socket joints with the arms to allow them to move freely.

2. To provide large areas of bone to which the powerful muscles of the arms can be attached.

THE PELVIC GIRDLE

The pelvic girdle (hip girdle) is fused to the backbone to form a strong shallow basin. It consists of two bones, a right and left pelvis (each formed from the fusion of 3 smaller bones).

Functions

1. To form ball and socket joints with the legs to make walking possible.

2. To provide large areas of bone to which the powerful muscles of the legs can be attached.

THE LIMBS

These are called **pentadactyl limbs** because they each have five **digits** (fingers or toes). Each limb consists of a long upper bone of the upper arm or thigh, 2 parallel lower bones of forearm or lower leg, small bones which form wrist or ankle, 5 bones which form the palm or foot, and fingers or toes which are jointed.

The **arms** perform a variety of movements. They can do this because of

1. The ball and socket joint at shoulder.
2. The hinge joint at elbow.
3. The rotation of the radius around the ulna enabling the hand to be turned over.
4. The gliding joints of the 8 small bones of the wrist (**carpals**) which allow the hand to rotate.
5. The ability of the thumb to rotate to a position opposite the fingers.
6. The jointed fingers.

The **legs** support the weight of the rest of the body and move it from place to place. They can do so because of:

1. The bones being generally larger and stronger.
2. The ball and socket joint at the hip.
3. The hinge joint at the knee (protected by the knee cap).
4. The seven ankle bones (**tarsals**) which interlock and form gliding joints to make a strong movable region.
5. The five **metatarsals** arranged in one plane to help in support and balance.

MUSCLES

MUSCLE TISSUE

Muscle is the tissue responsible for movement of the various parts of the body. It can do this because it is able to contract and then relax. There are three main types of muscle.

1. **Smooth Muscle** is found mainly in the walls of the internal organs, e.g. oesophagus, stomach, intestine, ureter, blood vessels. It consists of uninucleate cells which interlock (Fig. 3.7). Movement is controlled by the autonomic nervous system (p. 62) and not by the will. When a smooth muscle cell contracts, the pull is transmitted to adjacent cells which pass it on. In this way the force of contraction is transmitted uniformly through the muscle, causing, for example, peristalsis and the waves of contraction which pass along arteries.

Fig. 3.7 Smooth muscle tissue.

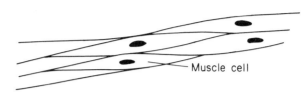

Muscle cell

2. **Cardiac Muscle** is heart muscle and consists of muscle cells which have a striped appearance. This too is under the control of the autonomic nervous system and not the will, and contracts automatically and rhythmically.

3. **Skeletal Muscle** is attached to bones. It can also be called **striped** or **striated muscle** as it has this appearance when seen under a microscope. As it is under the control of the will it is sometimes called **voluntary muscle.** The red colour of skeletal muscle is due to the presence of the pigment **myoglobin** which is similar to haemoglobin. This muscle tissue consists of bundles of parallel **muscle fibres** held together by connective tissue. Each fibre is multinucleate with a tough outer membrane containing cytoplasm and **fibrils** of protein. The fibrils extend along the whole length of the muscle fibre and under the microscope they all appear to have a similar pattern of alternate light and dark bands, thus giving the fibre a striped effect (Fig. 3.8). Nerve endings penetrate the fibres and cause them all to contract simultaneously. When they contract they become shorter and thicker, the stripes move closer together and a bone will be caused to move. The muscle fibres rarely extend the whole length of the muscle and they are more numerous in the middle giving the muscle the characteristic bulged appearance.

Fig. 3.8 Left: part of striped muscle fibre; Right: T.S. bundle of striped muscle fibres.

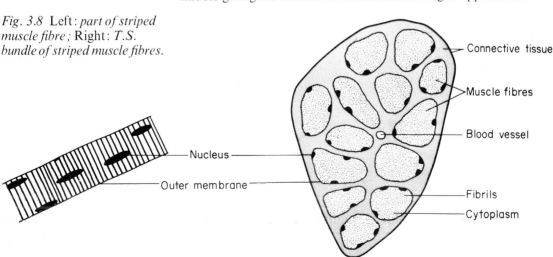

Connective tissue

Muscle fibres

Blood vessel

Nucleus

Outer membrane

Fibrils

Cytoplasm

TENDONS

Tendons at either end of the muscle consist of strong, white fibres and attach the muscle to bone by penetrating the periosteum. The end of the muscle attached to a bone that is stationary is called the **origin,** and the other end which is attached to the bone that moves is called the **insertion;** e.g. the biceps muscle has its origin in the scapula and insertion in the radius, the triceps muscle has its origins in the scapula and humerus and insertion in the ulna (Fig. 3.9).

Exercise can develop and enlarge muscles as it causes increase in the width of the individual muscle fibres. Muscle tissue also stores glycogen so that it has a source of food from which it can readily obtain energy when it is required to work.

HOW SKELETAL MUSCLES WORK

Movement of bones requires the co-ordinated contraction and relaxation of **antagonistic** (opposing) sets of muscles. For example, raising and lowering the fore-arm involves the **biceps muscle** and the **triceps muscle,** which are antagonistic to each other. When the biceps contracts the triceps relaxes and the fore-arm is raised; when the triceps contracts the biceps relaxes and the fore-arm is lowered (Fig. 3.9).

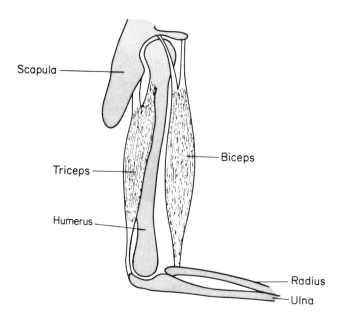

Fig. 3.9 Diagram to show an antagonistic pair of muscles— the triceps and biceps—which cause movement at the elbow joint.

22

Chapter Four

BLOOD AND ITS CIRCULATION

BLOOD

Blood is a red fluid which circulates continuously around the body. An adult contains about 4–5 litres (7–9 pints) of blood. It consists of:

Plasma	—55%
Red blood cells	
White blood cells	—45%
Platelets	

Plasma is the liquid part of the blood and is pale yellow in colour. It consists of:

Water—92%
Blood proteins, e.g. fibrinogen, antibodies (globulin), albumin.
Digested food substances, e.g. glucose, amino acids.
Waste substances, e.g. urea, carbon dioxide.
Ions of sodium, chloride, calcium, potassium.

Serum is plasma with the fibrinogen removed.

Red Blood Cells

Red blood cells (**red corpuscles** or **erythrocytes**) are minute bi-concave discs (Fig. 4.1.a), 1 mm^3 of blood containing about 5 million red cells. They are made in the bone marrow and each has a nucleus when formed, but this disintegrates before the red cell is released into the blood. The red cells possess **haemoglobin** which contains iron and gives the blood its red colour. In the lungs the haemoglobin combines with oxygen to form **oxyhaemoglobin.** In this way oxygen is carried around the body to the tissues and is released where there is an oxygen shortage. Oxygenated blood is a bright red colour; blood which has given up oxygen is darker red. A red cell normally lasts for about 120 days and when worn out it is destroyed in the liver or spleen, the iron being retained for further use.

Anaemia can be due to lack of iron in the diet which prevents the manufacture of haemoglobin, or to the inability of the body to manufacture enough red blood cells.

Fig. 4.1 Blood

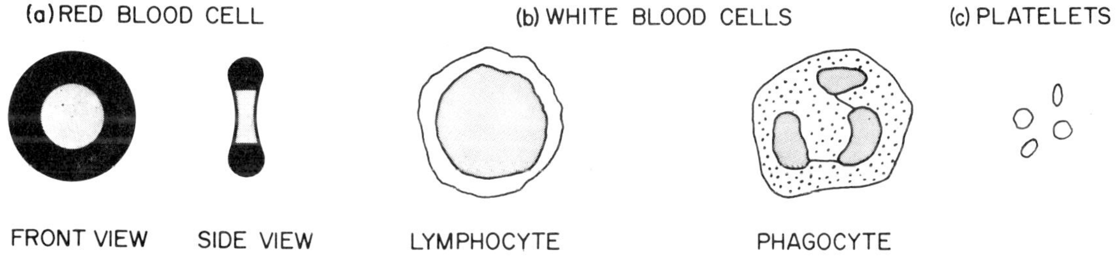

(a) RED BLOOD CELL

(b) WHITE BLOOD CELLS

(c) PLATELETS

FRONT VIEW   SIDE VIEW

LYMPHOCYTE

PHAGOCYTE

White Blood Cells

White blood cells (**white corpuscles** or **leucocytes**) are present in smaller numbers than red cells, 1 mm³ of blood containing about 7,000. Also they are larger, do not contain haemoglobin, are colourless, their shape is irregular and they possess nuclei of various shapes. They are made in the bone marrow and lymphatic tissue, and there are several different types including **lymphocytes** and **phagocytes** (Fig. 4.1.b). Their function is to protect the body. Phagocytes eat the germs, lymphocytes produce antibodies to destroy the germs or the toxins (poisons) they produce. The white cells can leave the blood stream and move around between the cells of the tissues in an amoeboid fashion. They collect at the site of infection and form pus which is the remains of dead bacteria, white blood cells and lymph.

Platelets

Platelets are small and their function is concerned with clotting (Fig. 4.1.c).

CLOTTING

Clotting of blood prevents excessive bleeding from minor wounds. When a blood vessel is damaged, the **fibrinogen** in the plasma is precipitated as fine threads of **fibrin.** The blood cells and platelets become caught in these tangled threads and a soft plug forms over the wound. This hardens into a scab, beneath which the tissue is repaired. Eventually the scab drops off leaving a scar.

Thrombosis is caused by a clot of blood called a thrombus forming within a blood vessel. Should a thrombus occur in one of the coronary arteries, the blood supply to the heart is interrupted and the condition is referred to as a **coronary thrombosis** or **'a coronary'.** Coronary thrombosis generally affects middle-aged or elderly people and it is thought that it may be due to stress combined with a fat-rich diet, insufficient exercise or smoking.

FUNCTIONS OF THE BLOOD

1. Oxygen is transported from the lungs to the tissues by the red blood cells.

2. Carbon dioxide is transported from the tissues to the lungs in solution in the plasma, or by the red blood cells.

3. Digested food is transported in the plasma from the small intestine to the tissues.

4. Urea is transported in the plasma from the liver to the kidneys to be excreted.

5. Hormones are transported in the plasma from the endocrine glands to the tissues they affect.

6. Heat produced in the more active parts of the body is distributed around the body by the blood.

7. Clotting prevents loss of blood.

8. The body is defended against disease by the white blood cells and clotting helps prevent the entry of germs through a wound.

BLOOD GROUPS

The blood a human being contains belongs to one of four blood groups:

Group A —approx. 40% of the population.
Group B —approx. 10% of the population.
Group AB—approx. 5% of the population.
Group O —approx. 45% of the population.

The different blood groups may or may not contain factors which cause red blood cells to agglutinate (clump together) when two different groups are mixed. If the wrong bloods are mixed together in a blood transfusion, the red cells will form clots which can block the smaller blood vessels and the effect on the patient can be fatal.

Provided that the blood has been cross-matched to exclude agglutination due to Rhesus factor incompatibility (see below) or rare blood factors,

A patient can always receive a blood transfusion of the same blood group as his own.

Group AB can receive blood from any other group (**universal recipients**).

Group O can give blood to any other group (**universal donors**).

To Test for Blood Group

Place a drop of serum from Group A and a drop of serum from Group B on a white tile. Then add a drop of the patient's blood to each:

If the patient's blood is Group A it will agglutinate in Serum B.
If the patient's blood is Group B it will agglutinate in Serum A.
If the patient's blood is Group AB it will agglutinate in both Serum A and B.
If the patient's blood is Group O it will not agglutinate in either.

RHESUS (Rh) FACTOR

This is another factor which may or may not be present in blood. Most of the population possess it and are said to be Rhesus positive (Rh+), the rest do not and are Rhesus negative (Rh−). People with Rh− blood should not be given Rh+ blood as it will cause them to produce an antibody against the Rh factor. One transfusion will not usually cause much harm but a second one could cause agglutination of the red blood cells which would be serious or fatal.

The Rh factor is important during pregnancy if the mother is Rh− and the father Rh+, as the child may be Rh+. If it is, and during development some of the embryo's blood seeps into the mother's circulation she will produce an antibody against the Rh+ factor. Should the mother's plasma, now containing the antibody, seep back into the embryo it will cause agglutination of the red blood cells and serious damage or stillbirth. If the child survives birth then most of its Rh+ blood is exchanged by transfusion for compatible blood, and this removes the offending antibodies. The first baby is rarely affected because the mother needs time to build up a dangerous quantity of antibodies. It is now possible to immunise Rh− mothers to prevent them from manufacturing the antibodies.

BLOOD BANKS

Blood banks are refrigerators in hospitals in which blood is stored in 500 cm³ bottles. It has been obtained from donors by the Blood Transfusion Service and can be stored for about a month if sodium citrate is added to prevent clotting.

Freeze-dried plasma and serum can be stored indefinitely and are used in cases of shock or burns when the patient requires an immediate increase in the volume of blood. New red cells will be produced rapidly once the volume is restored.

THE BLOOD VASCULAR SYSTEM

The blood vascular system is concerned with the circulation of blood around the body and consists of a heart, arteries, capillaries and veins.

THE HEART

The heart is a muscular pump situated in the thorax between and in front of the lungs. It is centrally placed, but tilted to the left, and this causes the heart beat to be felt on the left side of the body. It is about the size of a closed fist and is divided into four chambers (Fig. 4.2).

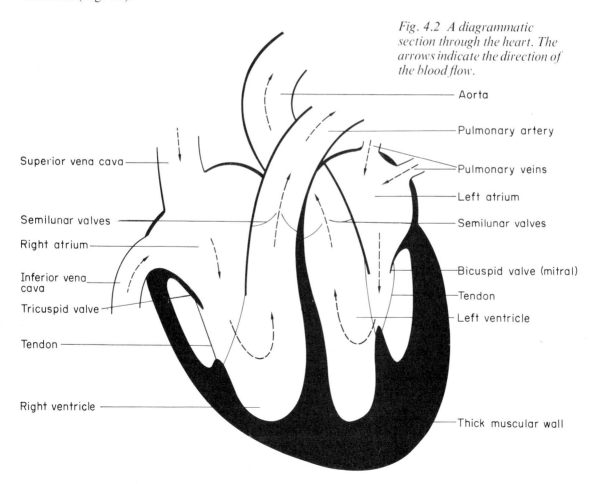

Fig. 4.2 A diagrammatic section through the heart. The arrows indicate the direction of the blood flow.

Aorta

Pulmonary artery

Superior vena cava

Pulmonary veins

Left atrium

Semilunar valves

Semilunar valves

Right atrium

Inferior vena cava

Bicuspid valve (mitral)

Tricuspid valve

Tendon

Tendon

Left ventricle

Right ventricle

Thick muscular wall

The Atria

The two upper thin-walled chambers which receive blood from the veins. (Each one is an **atrium.**)

The Ventricles

The two lower thick-walled chambers which pump blood into the arteries. The left ventricle has a thicker wall than the right ventricle as it has to pump blood further.

Heartbeat

This is caused by the alternate contraction and relaxation of the muscular walls of the heart. When the ventricles contract (the **systole**) the blood is forced out into the arteries, the **tricuspid** and **bicuspid** valves closing to prevent the blood returning to the atria. When the ventricles relax (the **diastole**) blood rushes in to fill them up, at the same time the **semilunar valves** close to prevent the return of the blood from the arteries. The rate of heart beat is about 70 times a minute in a sedentary adult. It is faster in a child and slower in old age. The rate increases with exercise up to 140 times a minute in order to supply the tissues with extra oxygen and food. It is also increased by nervous excitement, fright and disease.

Fig. 4.3 Diagram showing the main arteries of the body. Note: *the two coronary arteries which leave the aorta to take blood to the heart muscle are not shown.*

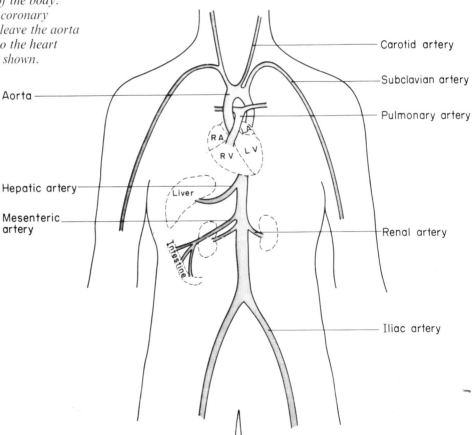

Pacemaker

A small mass of special tissue situated in the wall of the right atrium. This keeps the heart beating rhythmically.

ARTERIES

Arteries are blood vessels which carry blood away from the heart. The walls are muscular and blood is pumped through them by the heart. At each heartbeat a wave of blood is driven through the artery which distends the artery wall as it passes along and can be felt as the **pulse** wherever an artery of large or medium size passes near the surface of the body. The pulse rate is the same as the heartbeat. Fig. 4.3 shows the main arteries of the body. These branch repeatedly, the smaller, narrower branches being called **arterioles.**

VEINS

Veins are blood vessels which carry blood to the heart; the smaller veins being called **venules.** Fig. 4.4 shows the main veins of the body. The walls are thinner than those of the arteries (Fig. 4.5) and there are semilunar valves along the course of the vein to ensure that blood flows in one direction only. The valves remain open so long as the blood flows towards the heart, but if it moves in the opposite

Fig. 4.4 Diagram showing the main veins of the body.

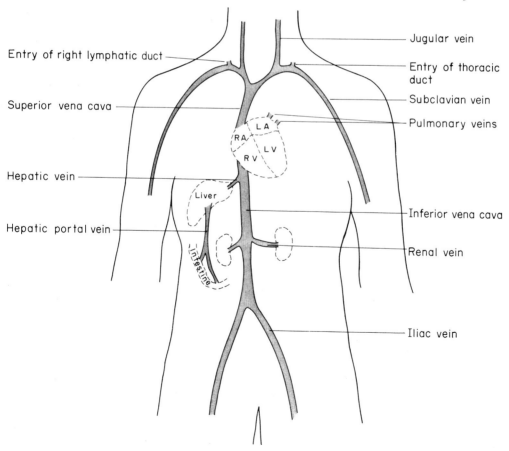

- Jugular vein
- Entry of right lymphatic duct
- Entry of thoracic duct
- Superior vena cava
- Subclavian vein
- Pulmonary veins
- L A
- R A
- L V
- R V
- Hepatic vein
- Liver
- Inferior vena cava
- Hepatic portal vein
- Renal vein
- Intestine
- Iliac vein

direction this closes the valve and prevents a backward flow (Fig. 4.5). The condition known as **varicose veins** occurs if the valves fail to close completely, the veins becoming swollen due to the increased volume of blood in them.

The flow of blood through the veins is due to the smaller veins becoming filled with blood from the capillaries, which are then subject to squeezing by contractions of the surrounding muscles. This squeezing tends to drive the blood in both directions, but the valves prevent a backward flow. The suction action of the heart also draws blood towards it, and the force of gravity encourages blood flow from the head and neck. Because the rate of blood flow through the veins is slower, they need to be larger and more numerous than arteries in order to return the necessary amount of blood to the heart.

Fig. 4.5 Blood vessels.

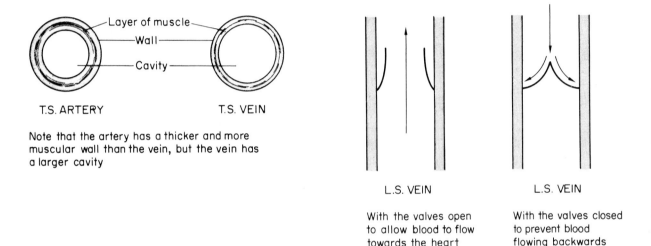

T.S. ARTERY   T.S. VEIN

Note that the artery has a thicker and more muscular wall than the vein, but the vein has a larger cavity

L.S. VEIN

With the valves open to allow blood to flow towards the heart

L.S. VEIN

With the valves closed to prevent blood flowing backwards

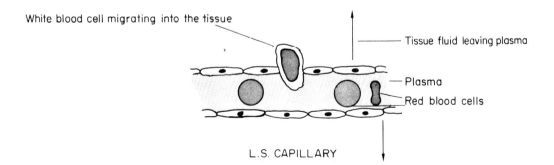

White blood cell migrating into the tissue

Tissue fluid leaving plasma

Plasma

Red blood cells

L.S. CAPILLARY

CAPILLARIES

Capillaries are the smallest blood vessels, allowing red blood cells to pass along in single file (Fig. 4.5). They link arterioles to venules and the walls are one cell thick. As the blood passes through the capillaries some of the liquid part of the plasma is forced through the walls and bathes the cells of the tissues. This is called **tissue fluid** and takes food and oxygen to the cells. Most of it returns direct to the blood system via the capillaries and venules. The rest drains into the lymphatic system and is returned to the blood system at two points near the heart.

CIRCULATION OF BLOOD

The blood circulates continuously around the body through the heart, arteries, capillaries and veins. The right side of the heart sends it to the lungs to collect oxygen. It then returns to the left side of the heart to be pumped out through the **aorta.** This gives off branches to the various organs and parts of the body. The branches repeatedly sub-divide until they are so small as to be capillaries. Blood from the capillaries is collected up by venules

Fig. 4.6 Diagram to show the general course of circulation.

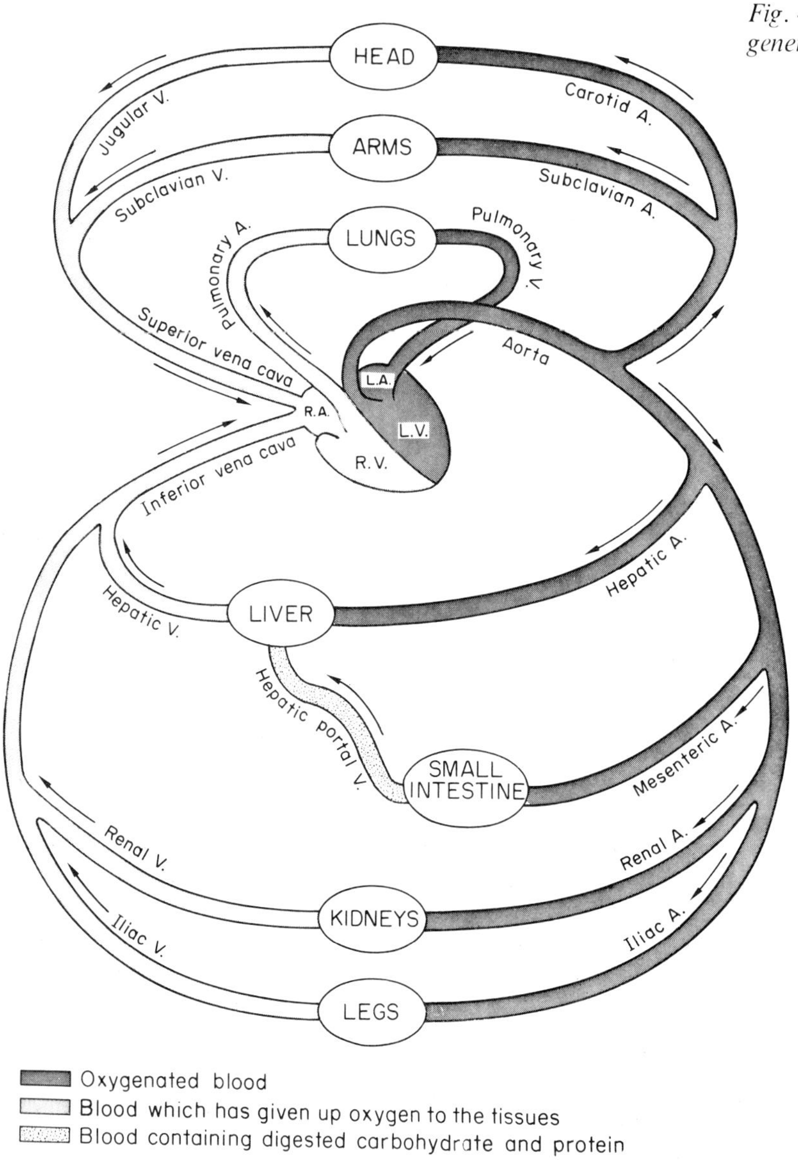

Oxygenated blood
Blood which has given up oxygen to the tissues
Blood containing digested carbohydrate and protein

which join together to form larger veins which return blood to the right side of the heart (Fig. 4.6).

Blood circulates around the body at a remarkable rate taking about 28 seconds to go from the right foot to the left foot. To do this it must pass through the right iliac vein, inferior vena cava, right atrium, right ventricle, pulmonary artery, capillaries of lung, pulmonary vein, left atrium, left ventricle, aorta and left iliac artery.

TISSUE FLUID

Tissue fluid is the liquid which leaves the blood and passes through the capillary walls to bathe the living cells. (The clear liquid which collects in a blister is tissue fluid.)

Tissue fluid contains:
1. Oxygen for tissue respiration.
2. Food for energy, growth, repair.
3. White blood cells to combat disease.
4. Carbon dioxide and other waste products from the cells.

THE LYMPHATIC SYSTEM

The lymphatic system is composed of vessels which collect up some of the tissue fluid and return it to the blood system. The smallest of these vessels are called lymphatic capillaries and they are thin-walled, blindly ending tubes forming a dense network in most of the tissues of the body. They collect tissue fluid which becomes called **lymph** as soon as it enters them. The lymphatic capillaries unite to form larger vessels.

The main lymphatic vessel is the **thoracic duct** which lies in front of the vertebral column. It collects lymph from the legs, left arm, left side of the thorax, and also from the abdomen. This includes lymph from the lacteals of the villi of the small intestine (Fig. 6.6). When fat droplets are present this lymph is a milky colour and is called **chyle.** The thoracic duct opens into the **left subclavian vein.** Lymph from the head, right arm and right side of the thorax is collected into the **right lymphatic duct** which opens into the **right subclavian vein.**

Lymph Nodes

Lymph nodes (**lymph glands**) are swellings along the course of the lymph vessels (rather like beads on a string). They consist of lymphatic tissue and occur in all parts of the body particularly at the armpit, side of the neck, elbow joint and in the groin. As lymph passes through the lymph node it is filtered through narrow passageways to purify it. Here white blood cells called phagocytes engulf bacteria and other foreign particles, and the lymphocytes produce antibodies to destroy or inactivate them. Lymph nodes manufacture lymphocytes and when infection occurs enormous numbers can be produced.

THE SPLEEN

The spleen is a dark red coloured organ lying just below the diaphragm and to the left of the stomach. It is approximately 11 cm long, 7 cm wide and 3 cm thick. It contains a great deal of lymphatic

31

tissue and its functions are:

1. To produce lymphocytes.
2. To produce antibodies.
3. To destroy worn out red blood cells and store the iron which is retained until it is required for the formation of more haemo-globin.
4. Act as a storage organ for a reserve supply of blood.

THE THYMUS GLAND

The thymus gland is an irregularly shaped gland lying in the thorax behind the breast bone. It is concerned with the production of lymphocytes and the formation of antibodies which help to combat disease. It is a very important gland in childhood but begins to decrease in size after puberty.

Patches of lymphatic tissue are found throughout the body includ-ing the **tonsils, adenoids** and **appendix.**

Chapter Five

RESPIRATION

Respiration is the process by which the body produces energy. It involves breathing in order that oxygen can be obtained and carbon dioxide excreted. The oxygen is required for the oxidation

Fig. 5.1 Diagram to illustrate respiration.

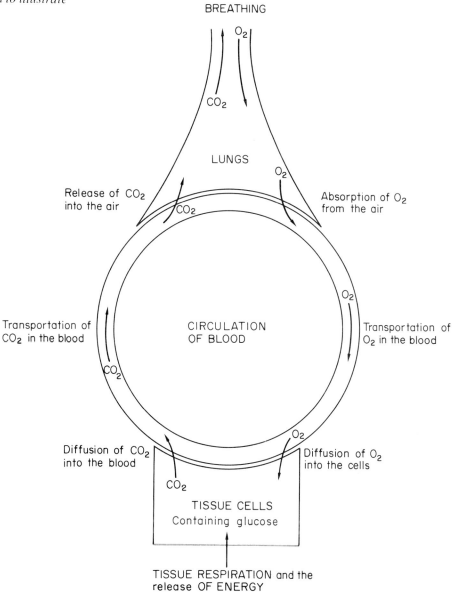

BREATHING

LUNGS

Release of CO_2 into the air

Absorption of O_2 from the air

Transportation of CO_2 in the blood

CIRCULATION OF BLOOD

Transportation of O_2 in the blood

Diffusion of CO_2 into the blood

Diffusion of O_2 into the cells

TISSUE CELLS
Containing glucose

TISSUE RESPIRATION and the release OF ENERGY

of food (mostly glucose) to release the stored energy it contains; carbon dioxide is given off as a waste product (Fig. 5.1).

There are five stages:
1. Breathing (also called **external respiration**).
2. Absorption of oxygen.
3. Transportation of oxygen from lungs to tissues.
4. Tissue respiration and the release of energy (also called **internal respiration**).
5. Elimination of the waste product carbon dioxide.

THE RESPIRATORY SYSTEM

The respiratory system is designed for breathing and consists of the respiratory tract, two lungs, thorax wall and diaphragm (Fig. 5.2).

Fig. 5.2 Diagram of a section through the head (side view) and thorax (front view) to illustrate the respiratory system.

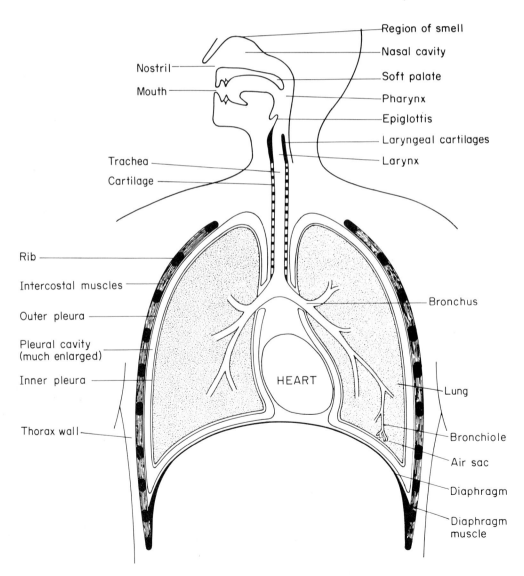

The Lungs

The lungs are two large spongy organs occupying the greater part of the thorax. They are connected to the adjacent heart by the pulmonary arteries and veins, and their sponginess is due to being composed of millions of minute air sacs. In young children they are a pale pink colour but in adults, especially city-dwellers, they become a grey colour because of the accumulation of inhaled dust.

The Pleurae

The inner and outer pleurae are the two membranes that surround each lung. The inner pleura is attached to the lung and the outer pleura to the thorax wall. It is very difficult to separate the pleurae, but between them, in the pleural cavity, is a thin film of fluid that allows them to move over one another easily during breathing.

The Respiratory Tract

This consists of hollow passages within the head and thorax along which air passes during breathing. The various regions are called nose, pharynx, larynx, trachea, bronchi and bronchioles. Most of these passages are lined with **ciliated mucous membrane** which produces **mucus** (a slimy secretion) to keep the passages moist and trap dust and bacteria. Within the lungs the cilia move the mucus to the bronchi and trachea where it is removed by coughing. 'Blowing the nose' and sneezing remove mucus from the nasal passages.

The Nose

Air enters the nose through the **nostrils** and as it passes along the narrow nasal passages it is warmed and moistened. Some of the dust and dirt becomes trapped in the mucus and chemical substances in the air are detected by special sensory cells which give the sense of smell.

The Pharynx (Throat)

This is at the junction of nose and mouth. Air can also enter the body through the mouth if the nose is blocked by excess mucus, as during a cold.

The Larynx (Voice Box or 'Adam's Apple')

This connects the pharynx with the trachea. It is about 5 cm long and is surrounded by cartilages—the **laryngeal cartilages.** The **vocal cords** are two bands of tissue attached to the sides of the larynx which can be tightened or slackened as required. This alters the size of the opening through which air is expelled from the lungs and allows the cords to vibrate so as to produce different sounds. In speech, these sounds are modified by the position of the tongue, lips and jaws.

The Trachea (Windpipe)

This is a tube about 12 cm long supported by 16–20 rings of cartilage. These not only keep the trachea open but also allow free movement of the head. The rings of cartilage are incomplete along the side of the trachea adjacent to the oesophagus and this allows the oesophagus to expand as solid food passes down to the stomach.

During swallowing, the **epiglottis** covers the entrance to the trachea, but if by any chance food does enter the windpipe, choking and coughing occur to clear the passage.

The trachea divides into two branches called **bronchi** (each one is a bronchus). These subdivide repeatedly, the smaller branches being called **bronchioles.** The walls are strengthened with cartilage to keep these tubes open.

The bronchioles terminate in **air sacs** which lack cartilage and have thin elastic walls. Each air sac gives rise to pouches called **alveoli** which are surrounded by a dense network of capillaries (Fig. 5.3).

Fig. 5.3 Diagram of an air sac.

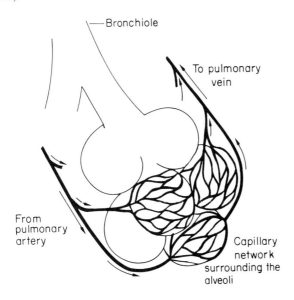

RESPIRATION

1. BREATHING

Breathing is the process of getting air into and out of the body. It is a rhythmical process, air is sucked into the body when the lungs are made to expand and it is pushed out of the body when the lungs are squeezed. During this process some of the air in the alveoli is exchanged for air from the outside.

2. ABSORPTION OF OXYGEN

The oxygen content of the air in the alveolus is greater than that in the blood of the surrounding capillaries. Hence oxygen dissolves in the moisture lining the alveolus wall, then diffuses through this wall, through the capillary wall and into the blood. Here it is taken up by the haemoglobin of a red blood cell to form oxyhaemoglobin (Fig. 5.4).

3. TRANSPORTATION OF OXYGEN

The oxygenated blood is collected up and returned by the pulmonary vein to the heart to be pumped out to all the tissues of the body. In the tissues the arteries subdivide many times to form capillaries which penetrate between the cells. If a cell has an oxygen shortage the oxygen will leave the haemoglobin and move by diffusion from the capillary into the cell.

Fig. 5.4 Diagram of a section through an alveolus and a capillary.

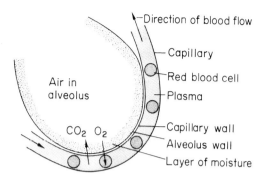

4. TISSUE RESPIRATION

Enzyme action within the cell results in the oxygen being used to oxidise glucose to carbon dioxide and water. The energy released by this process is used to convert one form of phosphate compound, **adenosine diphosphate (ADP),** into **adenosine triphosphate (ATP).** ATP is the fuel which is used, as required, to provide the energy for the various metabolic processes of the cell, and the heat energy essential to keep the body warm. When the energy stored in ATP has been used, it becomes converted back to ADP and phosphate. ATP is regenerated when more energy is released by oxidation.

Respiration can be summed up in the following chemical equation:

$$\underset{\substack{\text{glucose} \\ \text{containing} \\ \text{stored} \\ \text{energy}}}{C_6H_{12}O_6} + \underset{\text{oxygen}}{6O_2} \xrightarrow[\text{action}]{\text{enzyme}} \underset{\substack{\text{carbon} \\ \text{dioxide}}}{6CO_2} + \underset{\text{water}}{6H_2O} + \text{energy}$$

5. ELIMINATION OF CARBON DIOXIDE

Carbon dioxide is given off as a waste product of tissue respiration and needs to be excreted from the body. It diffuses from the cell into the capillary and is carried in the blood stream to the heart where it is pumped to the lungs. It travels in the blood mainly in the form of bicarbonates, most of which are in the plasma, but some are in the red blood cells. When the carbon dioxide reaches the lungs the concentration in the capillaries is greater than that in the air in the alveolus, so carbon dioxide diffuses from the blood through the walls and into the alveolus. It is then excreted from the body when the air is breathed out.

MECHANISM OF BREATHING

Inspiration (breathing in) occurs when the thoracic cavity enlarges because the diaphragm muscles contract to flatten the diaphragm and the intercostal muscles contract to pull the rib cage upwards and outwards. Because the pleurae are attached to the thorax wall and the lungs, and cannot be separated, when the thorax enlarges the lungs expand and create space. Air rushes in through the nostrils and respiratory tract to fill up this space (see Fig. 5.5).

Expiration (breathing out) occurs when the muscles of the diaphragm relax, allowing it to rise, and the intercostal muscles relax allowing the rib cage to move downwards and inwards. These movements reduce the space within the thoracic cavity and air is squeezed out.

Breathing is a regular and automatic process under the control of the nervous system. Carbon dioxide is the most important factor governing breathing and an increase in the CO_2 content of the alveolar air causes a similar increase in the blood which stimulates the nervous system to increase the rate and depth of breathing (p. 80).

Fig. 5.5 Diagrams to illustrate the mechanism of breathing.

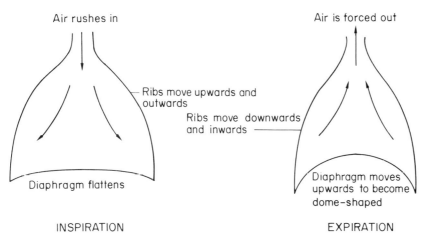

INSPIRATION

EXPIRATION

Lung Capacity

The lungs always contain a certain amount of air, the **residual air,** even after expiration. About 500 cm³ of **tidal air** passes in and out during quiet breathing. After strenuous exercise or during forced breathing the amount passing in and out increases up to a maximum of about 4000 cm³, this is known as the **vital capacity** of the lungs.

Composition of Respired Air

	AIR BREATHED IN	AIR BREATHED OUT
Nitrogen	79% approx.	79% approx.
Oxygen	21% approx.	17% approx.
Carbon dioxide	0·03% approx.	4% approx.
Water vapour	varies	saturated.

N.B. Approximately 400 cm³ of water is lost daily from the body by breathing.

ARTIFICIAL RESPIRATION

The **mouth-to-mouth** method is by far the best way to try to revive a person who has stopped breathing as a result of drowning, electric shock, suffocation, etc. It is most important to act *quickly* as damage to the brain cells occurs if they are deprived of oxygen for more than 4–6 minutes.

1. Place the patient on his back, if possible with the head at a lower level than the stomach as this helps diaphragm movement.

2. Remove any food or false teeth from the mouth and throat.

3. Tilt the head backwards as far as possible to prevent the tongue from blocking the airway.

4. Push the jaw forward to open up the windpipe.

5. Close the patient's nose by pinching the nostrils, take a deep breath, place your mouth firmly over the patient's (a handkerchief can be placed in between), and blow into his mouth until the chest is seen to rise.

6. Quickly take your mouth away to allow the patient to breathe out.

7. Inflate the patient's lungs every six seconds, and continue to try to revive the patient for at least one hour.

8. With infants it is possible to place your mouth over both mouth and nose. Little puffs only should be blown into the infant's lungs at the rate of about 30 a minute, as too much pressure can cause damage to the lung tissue.

Chapter Six

DIGESTION

Digestion is the process by which food is broken down with the aid of enzymes into simpler soluble substances which can be absorbed into the body.

Of the six necessary food substances (p. 98), generally only carbohydrates, fats, and proteins need to be digested. The other three— vitamins, minerals and water do not need to be digested as they are already in the right state to be absorbed.

Enzymes are the chemical substances that activate all living processes. **Intracellular enzymes** occur inside cells and control metabolism. **Extracellular enzymes** are produced by cells but have their effect outside the cells, e.g. digestive enzymes.

The **digestive enzymes** speed up the rate at which large particles of food are broken down into smaller ones. These enzymes work best at about body temperature and the digestive juice of the stomach is acid, whilst the digestive juices of the intestine are alkaline.

CHARACTERISTICS OF ENZYMES

1. They are made of protein.
2. They are formed by living cells.
3. They act as **catalysts,** that is, they speed up the rate of chemical change. Although there are a large number of different enzymes each acts only on a particular substance or group of substances, causing a specific chemical change, and does not affect any others.
4. They act most rapidly at the correct pH*—either acid or alkaline depending on the particular enzyme. An incorrect pH either slows down or stops enzyme action.
5. They work best within a narrow temperature range and they are destroyed by heat.

THE DIGESTIVE SYSTEM

This is designed for eating, swallowing, digesting and absorbing food and the elimination of unabsorbed matter from the body. It consists of teeth, mouth, pharynx, oesophagus, stomach, intestines, anus, salivary glands, liver and pancreas.

TEETH

The hard structures in the mouth for biting and chewing food.

Structure

Each tooth has a part called the **root** embedded in the jawbone and a part called the **crown** protruding into the mouth, the **neck** being the region where they meet (Fig. 6.1). The tooth is held in

* The pH scale is used to express the degree of acidity or alkalinity of a solution; pH 7 is neutral, less than 7 is acid, more than 7 is alkaline.

position by fibrous tissue which forms a firm connection between root and jaw-bone. Each tooth is composed of four substances.

1. *Dentine* forms the greater part of the tooth and is a hard bone-like tissue. It is penetrated by minute tubules radiating from the pulp cavity, each containing a protoplasmic fibre which keeps the tissue alive. Dentine is sensitive to touch, temperature, acids, sugars, etc., because the protoplasmic fibres transmit the information to the nerves within the pulp cavity.

2. *Pulp* is the soft matter in the cavity in the centre of the tooth and contains blood vessels and nerve endings sensitive to pain, indicating when the tooth needs attention.

3. *Enamel* covers the crown of the tooth. It is the hardest substance in the body, protecting the tooth and forming a hard, biting surface. If it is chipped or worn away by bacterial acids the dentine is exposed. The dentine is then liable to decay and if the nerves in the pulp cavity are exposed this causes toothache.

4. *Cement* is a thin layer of bone tissue which covers the root of the tooth.

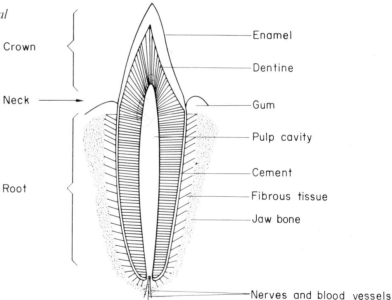

Fig. 6.1 Diagram of a vertical section through a tooth.

Crown

Neck

Root

Enamel

Dentine

Gum

Pulp cavity

Cement

Fibrous tissue

Jaw bone

Nerves and blood vessels

There are four types of teeth which vary in shape according to their function (Fig. 6.2).

1. Incisors are situated in the front of the mouth and have chisel-like edges for biting off pieces of food.

2. Canines are somewhat pointed and can be used for tearing off pieces of food. They are situated next to the incisors.

3. Premolars are used for grinding food into small pieces and they have larger, flatter surfaces. They occur next to the canines at the sides of the mouth.

4. Molars are larger than the premolars, but similar in form and function. They are not present in the milk teeth, and even in the permanent teeth the end molars, called the 'wisdom teeth', may take

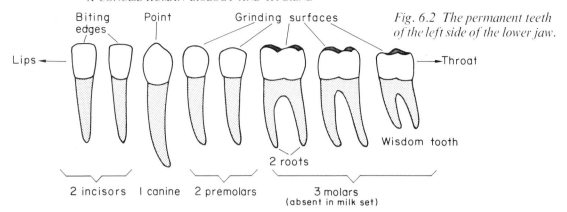

Fig. 6.2 *The permanent teeth of the left side of the lower jaw.*

a long while to develop. The upper molars have three roots, the lower molars have two.

Man, like most other mammals has two sets of teeth. The first set are called the **milk teeth** and they begin to appear in a definite order when the child is a few months old. There are twenty milk teeth, ten in each jaw—4 incisors, two canines and 4 premolars.

From the age of five onwards they are gradually replaced by the **permanent teeth** which have usually all appeared by the age of seventeen, although the wisdom teeth may be delayed. Besides replacing the milk teeth with the same type of tooth there are also 3 molars on each side of both jaws, making a total of 32.

THE ALIMENTARY CANAL

The alimentary canal is the passage in the body along which food passes so that it can be digested and absorbed (Fig. 6.3). Basically it is a muscular tube, lined with mucous membrane, extending from lips to anus with different parts being adapted for different functions to give the mouth, pharynx, oesophagus, stomach, small intestine, large intestine, rectum and anus. Digestive juices are either secreted by the glands of the lining of the canal or poured into the canal through ducts from special glands outside. As the food passes through the alimentary canal, it undergoes complex mechanical and chemical changes until it is in a suitable state to be absorbed. The indigestible residue is expelled through the anus.

Peristalsis

Peristalsis is the name given to the waves of muscular contraction which pass along the alimentary canal and drive food through it.

DIGESTION

DIGESTION IN THE MOUTH

The food is chewed up by the teeth into smaller pieces and mixed with saliva.

Saliva

Saliva is a watery fluid secreted by three pairs of glands opening by ducts into the mouth. It contains the enzyme **ptyalin** which begins the digestion of starch by splitting it up into the disaccharide, maltose. Saliva also moistens dry food enabling it to be tasted and swallowed.

Swallowing

After the food has been chewed for a short while, the lips close and the tongue pushes it to the back of the mouth. At the same time the soft palate closes the entrance to the nose, the epiglottis closes the entrance to the trachea, and the muscles of the pharynx push the food into the **oesophagus** (food pipe). Peristalsis moves the food through the oesophagus and the **cardiac sphincter** opens to allow it to enter the stomach.

DIGESTION IN THE STOMACH

The stomach is a muscular bag which stretches as it fills with food. Muscles of the stomach wall become active to churn up the food and mix it with **gastric juice** secreted by glands in the wall. Gastric juice contains the enzymes **pepsin** and **rennin,** as well as **hydrochloric acid.** The hydrochloric acid makes the stomach contents acid which allows the stomach enzymes to act, kills germs, and gradually stops the action of ptyalin.

Pepsin is the most important enzyme; it begins the digestion of proteins by splitting them up into peptides. Rennin clots milk and is important mainly in infancy.

Fig. 6.3 Diagram of the alimentary canal and associated glands.

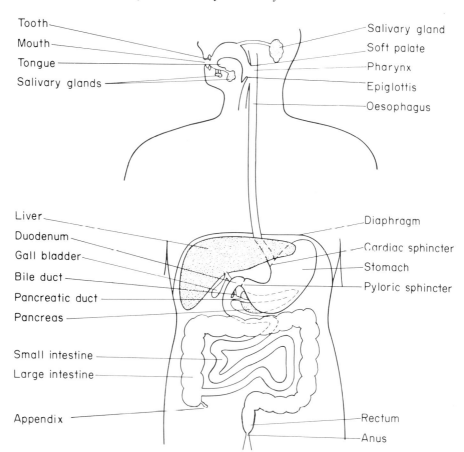

Tooth
Mouth
Tongue
Salivary glands

Salivary gland
Soft palate
Pharynx
Epiglottis
Oesophagus

Liver
Duodenum
Gall bladder
Bile duct
Pancreatic duct
Pancreas
Small intestine
Large intestine
Appendix

Diaphragm
Cardiac sphincter
Stomach
Pyloric sphincter

Rectum
Anus

The length of time the food stays in the stomach depends on the amount and type present. Carbohydrate foods pass quickly through, protein foods are kept for a longer time and fatty foods pass through most slowly of all. The churning action and gastric juice turn the stomach contents into a semi-liquid state called **chyme.** From time to time the **pyloric sphincter** opens to allow a little chyme through into the small intestine.

Fig. 6.4 Summary of diges- tion. Note: *the digestive juices and enzymes are shown in capital letters.*

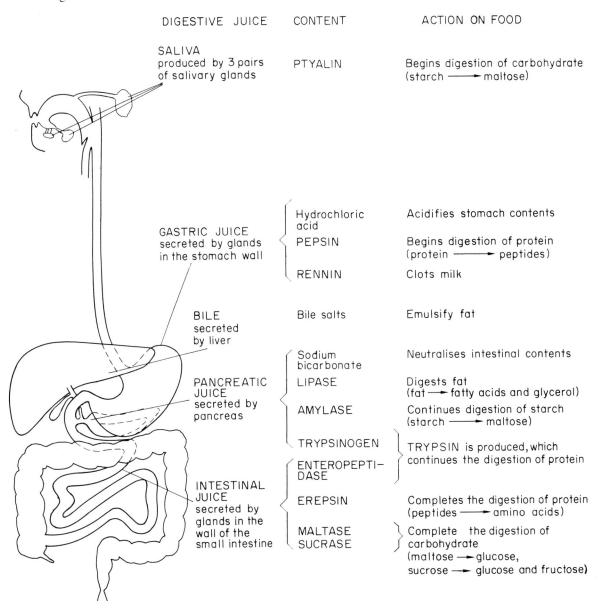

DIGESTIVE JUICE	CONTENT	ACTION ON FOOD
SALIVA produced by 3 pairs of salivary glands	PTYALIN	Begins digestion of carbohydrate (starch ——→ maltose)
GASTRIC JUICE secreted by glands in the stomach wall	Hydrochloric acid	Acidifies stomach contents
	PEPSIN	Begins digestion of protein (protein ——→ peptides)
	RENNIN	Clots milk
BILE secreted by liver	Bile salts	Emulsify fat
PANCREATIC JUICE secreted by pancreas	Sodium bicarbonate	Neutralises intestinal contents
	LIPASE	Digests fat (fat ——→ fatty acids and glycerol)
	AMYLASE	Continues digestion of starch (starch ——→ maltose)
	TRYPSINOGEN	TRYPSIN is produced, which continues the digestion of protein
	ENTEROPEPTI- DASE	
INTESTINAL JUICE secreted by glands in the wall of the small intestine	EREPSIN	Completes the digestion of protein (peptides ——→ amino acids)
	MALTASE SUCRASE	Complete the digestion of carbohydrate (maltose ——→ glucose, sucrose ——→ glucose and fructose)

DIGESTION IN THE SMALL INTESTINE

The small intestine is a long coiled tube in which the main part of digestion takes place. The first part is called the **duodenum** and the **bile duct** and **pancreatic duct** open into it. As the food moves through the intestine by peristalsis it is mixed with bile, pancreatic juice and intestinal juice.

Bile

Bile is a secretion of the liver and it is stored in the gall bladder until required. It is a yellowish-green colour and although it does not possess any enzymes it contains bile salts which emulsify fats, that is, split them up into minute droplets.

Pancreatic Juice

Pancreatic juice is secreted by the pancreas and contains **sodium bicarbonate** which neutralizes the acid chyme to allow a variety of enzymes to act on the food. Pancreatic juice contains three enzymes.

Amylase resumes the digestion of starch to maltose.

Lipase splits up fat into fatty acids and glycerol.

Trypsinogen which is converted to **trypsin** by enteropeptidase. Trypsin continues the digestion of protein and peptides to break them down to amino-acids.

Intestinal Juice

Intestinal juice is secreted by glands in the wall of the small intestine and contains several enzymes, including:

Enteropeptidase (formerly called **enterokinase**) which converts trypsinogen to trypsin.

Erepsin which splits peptides into amino-acids to complete the digestion of protein.

Maltase which splits up maltose into glucose to complete the digestion of carbohydrate.

Sucrase which splits up sucrose into glucose and fructose.

THE LARGE INTESTINE (COLON)

The material which passes from the small intestine into the large intestine is mainly water and undigested matter such as cellulose. It is moved along very slowly so that most of the water has time to be absorbed. The remains, called **faeces,** become more or less solid and are expelled at intervals through the **rectum** and **anus.** Enormous numbers of bacteria live in the large intestine and their presence has advantages, such as producing Vitamin K, and disadvantages, such as being a possible source of disease, as vast numbers of these bacteria are expelled with the faeces.

ABSORPTION

When the food has been digested and made soluble it is absorbed into the body through the intestinal wall.

Carbohydrate is absorbed mainly as glucose into the capillaries.

Protein is absorbed as amino-acids into the capillaries.

Fat is absorbed as fatty acids and glycerol which reunite to form fat droplets before passing into the lacteal. In some cases emulsified fat droplets can be absorbed without having to be first digested by lipase.

Absorption takes place as the food moves through the small intestine. It is about 5 metres long and the internal surface consists of a very large number of projections called **villi** and this greatly increases the area through which absorption can take place (Fig. 6.5). Each villus has a very thin wall and contains capillaries and a lacteal (Fig. 6.6). This means that the contents of the small intestine come very close to the blood system, merely being separated by very thin membranes across which the food substances can easily diffuse. The capillaries join up with the **hepatic portal vein** which takes the blood from the intestine to the liver. The **lacteals** are part of the lymphatic system and join up with the blood stream near the heart (p. 31).

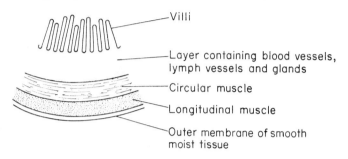

Villi

Layer containing blood vessels, lymph vessels and glands

Circular muscle

Longitudinal muscle

Outer membrane of smooth moist tissue

Fig. 6.5 Section through the wall of the intestine.

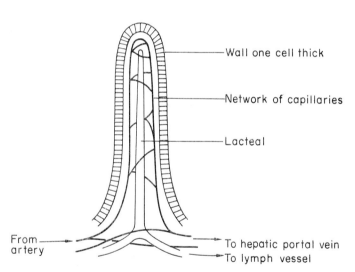

Wall one cell thick

Network of capillaries

Lacteal

From artery

To hepatic portal vein
To lymph vessel

Fig. 6.6 L.S. villus.

ASSIMILATION

After the food has been absorbed it is transported by the blood to the tissues. It is assimilated into the body when it is taken up by the cells of the tissues.

N.B. Alcohol is assimilated very rapidly if it is taken on an empty stomach. It passes through quickly to the small intestine where it is rapidly absorbed and soon afterwards is assimilated by the nervous system. If the stomach contains food, alcohol is retained there for a longer time, hence the rate of absorption is slower and the effect on the body is less, owing to the fact that the concentration of alcohol in the tissues is likely to be at a lower level.

THE LIVER

The liver is the largest gland in the body, weighing about $1\frac{1}{2}$ kg. It is dark-brown in colour with a smooth surface and is divided into five lobes. It is situated in the upper right side of the abdomen and is dome-shaped to fit under the diaphragm.

Four vessels are connected to it.

1. The hepatic artery which brings oxygenated blood from the heart.

2. The hepatic portal vein which brings blood containing sugar and amino-acids from the small intestine.

3. The hepatic vein which returns blood to the heart.

4. The bile duct which takes bile to the duodenum.

Functions

The liver has a large number of functions, many of which are concerned with helping to regulate the composition of the blood by keeping the various components at a constant level.

1. It maintains the level of the blood sugar. All the sugar absorbed by the intestine is taken to the liver. Here excess sugar in the blood is removed and stored as glycogen until it is needed. It is then gradually converted back to glucose to keep the amount of sugar in the blood at a constant level. When the hepatic portal vein brings more sugar from the next meal, the excess will again be stored as glycogen.

2. It breaks down excess amino-acids by a process called **deamination.** This is because amino-acids cannot be stored in the body, and those which are not required are deaminated, the nitrogen being removed in the form of ammonia. This is immediately converted to urea and transported to the kidneys to be excreted. The remains of the amino-acid (carbon, hydrogen and oxygen) are now used for energy or stored as glycogen or fat.

3. Removes fat from the blood and changes it so that it can either be used for energy or sent to fat tissue for storage.

4. It manufactures fibrinogen.

5. It stores iron and vitamins A, D, and B_{12}.

6. It produces bile which

 (a) assists digestion of fats.

 (b) excretes waste material from the haemoglobin of worn out red blood cells, thus giving it a greenish colour.

7. **Detoxication** The liver deals with poisonous substances which are either produced by the body or taken as drugs. It renders them harmless and gets rid of the waste via the kidneys or bile.

8. It generates heat as it is such a large and active organ. This helps to maintain the body temperature.

THE PANCREAS

The pancreas is a whitish gland about 20–25 cm long. It is in the abdomen situated behind the lower part of the stomach and is attached to the duodenum by the pancreatic duct.

Functions

1. Produces pancreatic juice containing enzymes which flows along the pancreatic duct to the duodenum.

2. Produces the hormone insulin in special groups of cells called Islets of Langerhans. This is carried round the body in the blood stream and helps to control the carbohydrate metabolism of the body (p. 79).

ENZYME TESTS

N.B. A water bath (or beaker containing water), maintained at a temperature of about 37 °C (98·4 °F), should be used for these experiments. The enzymes work most rapidly at this temperature.

EXPERIMENT TO DEMONSTRATE THE ACTION OF THE ENZYME PTYALIN

Ptyalin is also called **salivary amylase.** This experiment also applies to **amylase,** and **diastase** (the plant enzyme which converts starch to sugar).

Method

Mix a very small quantity of flour with some water and pour a little into two test-tubes labelled A and B. Add a drop of iodine to each test-tube and this turns the contents a blue-black colour as flour contains starch. To test-tube A add some saliva as this contains the enzyme ptyalin.

Result

After a while the contents of the tube A gradually turned from black to grey to white. The contents of tube B remained black.

Conclusion

Saliva caused the disappearance of the starch from tube A due to the enzyme ptyalin. Benedict's test (p. 104), applied to tubes A and B at the end of the experiment will show that sugar is present in A but not in B. Hence ptyalin causes the change from starch to sugar.

If the saliva had been boiled before adding it to tube A, the contents would have remained black as enzymes are destroyed by heat.

EXPERIMENT TO DEMONSTRATE THE ACTION OF PEPSIN ON PROTEIN

Method

As egg-white contains protein (p. 99), put a little fresh egg-white in a test-tube, add three times as much water and shake well. Heat well to coagulate the protein. Then strain through muslin to remove the lumps. Dilute this cloudy protein mixture with water and pour a little into four test-tubes labelled A, B, C, and D.

To tube A add a little dilute hydrochloric acid.

To tube B add a little pepsin powder.

To tube C add both dilute hydrochloric acid and pepsin powder.

Leave tube D as a control.

Result

After leaving for a while in a warm place, the contents of tube C turned from cloudy to clear, the others all remained cloudy.

Conclusion

The enzyme pepsin required an acid medium in which to act on the protein. Had the contents of tube C been boiled the enzyme would have been destroyed and the contents would have remained cloudy.

EXPERIMENT TO DEMONSTRATE THE ACTION OF LIPASE ON FAT
Method

Use liquid fat in the form of olive oil or cooking oil and put a little into two test-tubes containing 50% alcohol and labelled A and B. Shake well, then test with litmus to show that the contents of the tubes are neutral. Add a little lipase powder to tube A, shake well and leave for a few hours.

Results

When tested with litmus the contents of tube A had turned acid whilst those of tube B had remained neutral.

Conclusion

Lipase had acted on the fat to produce fatty acids. To show that glycerol is also present, add a little weak copper sulphate solution and then a few drops of caustic potash solution. The deep blue colour will indicate the presence of glycerol.

Chapter Seven

EXCRETION

Excretion is the process by which the body gets rid of the waste products of metabolism.

The waste products are carried by the blood from all living cells to the organs of excretion which then eliminate them from the body. If these waste products were allowed to accumulate they would be poisonous to the body.

Organs of Excretion

1. Lungs excrete CO_2 in expired air (see Chap. 5).
2. Kidneys excrete urea and other substances in urine.
3. Skin excretes a small amount of urea and other substances in sweat.
4. Liver excretes the waste material from the haemoglobin of worn-out red blood cells in bile.

THE URINARY SYSTEM

The urinary system is concerned with the excretion of urine and consists of two kidneys, ureters, renal arteries, renal veins, and a bladder and urethra (Fig. 7.1).

Fig. 7.1 Diagram of the urinary system.

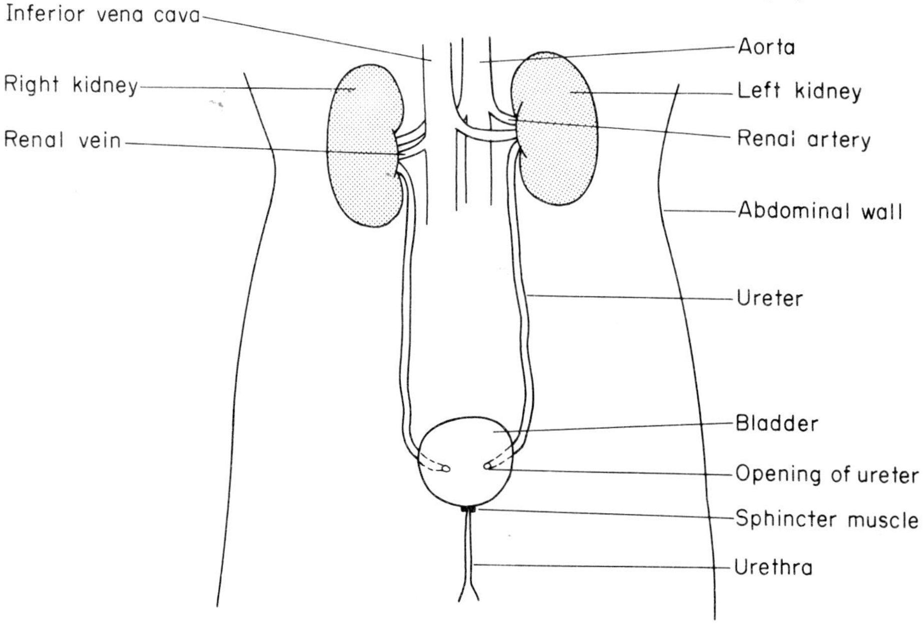

Inferior vena cava

Right kidney

Renal vein

Aorta

Left kidney

Renal artery

Abdominal wall

Ureter

Bladder

Opening of ureter

Sphincter muscle

Urethra

KIDNEYS
Functions
1. Excretion.
2. Water regulation.

The kidneys are able to perform these functions because:

1. They are situated near the heart and therefore in a position to receive large quantities of blood, about 1200 cm³ (2 pints) entering the kidneys every minute.

2. The blood is pumped into the kidney via the renal artery, the waste materials and excess water are removed, then the blood is returned into circulation via the renal vein.

3. Each kidney consists of a mass of tubules called **nephrons** surrounded by a vast network of capillaries.

4. The waste materials and excess water, known as **urine,** are collected from the nephrons and excreted via ureter, bladder and urethra.

Position
The kidneys are attached to the dorsal part of the abdominal wall, one on either side of the vertebral column, extending from the 12th thoracic to the 3rd lumbar vertebrae.

Structure
Each kidney is dark-red, bean-shaped, about 12 cm long, 6 cm wide, 2·5 cm thick, and weighs about 150 g. It is contained in a tough membrane and embedded in fat for protection. The left kidney is slightly larger and placed slightly higher than the right. Three vessels are connected to the kidney.

The **renal artery** conveys blood from the aorta to the kidney where it is filtered by the nephrons.

Fig. 7.2 L.S. kidney.

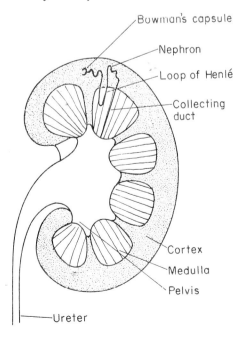

The **renal vein** returns the filtered blood to the inferior vena cava.
The **ureter** conveys urine from the kidney to the bladder.

A longitudinal section through the kidney shows clearly its three regions (Fig. 7.2).

Cortex—the darker outer region.

Medulla—the paler middle region which is scalloped-shaped where it meets the cortex.

Pelvis—the white inner region forming a branched cavity which narrows into the ureter.

The cortex and medulla consist of an enormous number of intertwined nephrons. The blind end of the nephron lies in the cortex and is expanded to form a cup-shaped capsule, called **Bowman's capsule,** which encloses a small knot of capillaries, called the **glomerulus.** The tubule leaving Bowman's capsule coils, then loops down into the medulla to form the **loop of Henlé,** then returns to the cortex and coils again before joining the collecting duct which leads to the pelvis of the kidney. The tubule wall is one cell thick and is surrounded by capillaries.

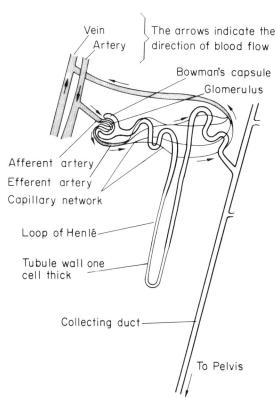

Vein
Artery
The arrows indicate the direction of blood flow
Bowman's capsule
Glomerulus
Afferent artery
Efferent artery
Capillary network
Loop of Henlé
Tubule wall one cell thick
Collecting duct
To Pelvis

Fig. 7.3 Diagram of a nephron.

HOW THE KIDNEY WORKS

In the glomerulus the force of the blood pressure is so great that, by a sieve-like action, much of the plasma is squeezed through the capillary wall, through the capsule wall and into Bowman's capsule.

This filtrate contains water, glucose, amino acids, urea, other nitrogenous waste and minerals. (The red blood cells, white blood cells, platelets and proteins are too large to pass through the walls.) As the filtrate passes along the tubule, selective re-absorption takes place. All the glucose and amino acids, much of the water and some of the minerals are re-absorbed back into the capillaries surrounding the tubule to restore the blood plasma to its correct composition. The fluid which is left is called **urine** and contains unwanted water, excess minerals, urea and other nitrogenous waste. The urine passes from the tubule, into the collecting duct, into the pelvis, and then peristalsis moves it down the ureter to the bladder.

THE BLADDER

The bladder is a muscular sac for storing urine. As urine drips into it, the bladder muscles relax and it swells as it becomes full. The outlet to the bladder is closed by the **sphincter muscle.***

At intervals this muscle relaxes, the bladder muscles contract and urine passes through the outlet to be expelled from the body through the urethra.

THE URETHRA

The urethra is a duct through which urine passes from the bladder to the outside of the body. It opens at the end of the penis in the male and the vulva in the female. (In the male, the urethra also serves as a duct for the passage of semen.) **Urination** is the action of eliminating urine from the body.

COMPOSITION OF URINE

Even in a healthy person the composition of urine varies widely but on average $100 \, cm^3$ urine contains, besides water,
2·0 g urea,
0·2 g other nitrogenous waste,
0·6 g sodium,
0·6 g chloride,
0·15 g potassium,
and very small amounts of other minerals.

If the urine contains glucose this may indicate sugar diabetes. However, it should be noted that sugar is sometimes excreted in the urine after a heavy meal. If blood or protein is present this may indicate kidney or bladder disease.

QUANTITY OF URINE

The average quantity of urine passed daily is about $1500 \, cm^3$. But the amount varies considerably according to:
1. Fluid intake.
2. Type of food eaten.
3. Amount of exercise taken and therefore amount of water lost in sweat and respiration.
4. Weather conditions, which will affect the amount of water lost in evaporation.

---

* A **sphincter** is a ring of muscle that closes an opening.

THE SKIN

Skin covers the surface of the body and also lines the mouth and rectum. Special outgrowths of the skin are hair and nails.

FUNCTIONS

1. To protect the underlying tissues.
2. To form a barrier against the entry of germs.
3. To prevent water loss.
4. To regulate body temperature and help keep it constant at about 37 °C (98·4 °F).
5. To contain nerve endings sensitive to touch, temperature and pain etc., which receive sensory impulses from the environment.
6. To excrete water, urea and minerals in the form of sweat.
7. To manufacture Vitamin D when exposed to sunlight.
8. To produce melanin to protect against sunburn.

STRUCTURE

The skin is composed of two layers—the epidermis and dermis (Fig. 7.4).

Fig. 7.4 Diagram of a section through the skin.

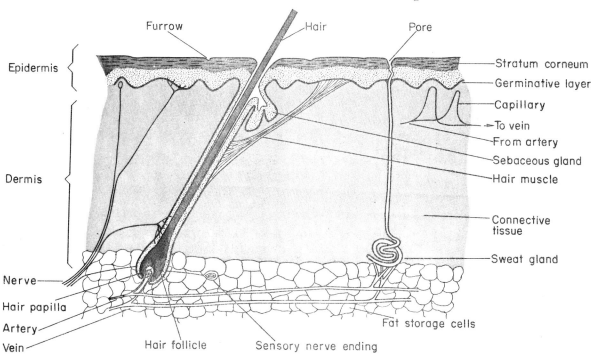

The Epidermis

The epidermis is the outer layer of skin and lacks blood vessels. The innermost cells form the **germinative layer (Malpighian layer).** These cells are constantly dividing to produce new cells which are then pushed outwards to make room for the newer cells being formed underneath. As they move outwards they retreat further from the blood supply and die through lack of food and oxygen. The cytoplasm becomes converted into the tough protein, **keratin,** which forms a distinct layer called the **stratum corneum** on the outside of the body. This protects the underlying tissues and it is renewed as fast as it is rubbed off. It also prevents water loss from the body. If too great an area is destroyed by burning, excessive water loss can cause death. Even a minor burn results in a blister containing a large amount of fluid.

The epidermis is thickest on the soles of the feet and palms of the hands which are the parts of the skin most liable to wear. It is thinnest on the lips.

There are special cells amongst the innermost epidermal cells which produce the dark pigment **melanin.** This pigment is passed to other epidermal cells giving colour to the skin and hair. All races have about the same number of pigment-producing cells in the skin but they are much more active in those people with dark skins. Sunlight also causes the pigment-producing cells to work harder and produce more melanin for protection against the sun's rays. Albinos are unable to produce melanin.

The germinative layer projects into the dermis to produce hair follicles, sebaceous glands and sweat glands.

The Dermis

The dermis is the innermost layer of skin. It is usually thicker than the epidermis and contains

Blood vessels
Hair follicles
Sebaceous glands
Hair muscles
Sweat glands
Nerves and nerve endings
Connective tissue
Fatty tissue

Blood Vessels

The dermis needs a plentiful supply of blood to circulate through the numerous capillaries, not only to keep the tissues alive, but also to assist in the regulation of body temperature (p. 57).

Hair Follicles

These are pits in the skin and each gives rise to a hair. The hair papilla at the base of the pit is well supplied with blood vessels. The hair papilla is surrounded by a group of germinative cells which constantly divide to form new cells which are added to the base of the hair causing it to grow. These are soon turned into the dead tissue keratin and consequently no pain is felt when the hair is cut. Growth of hair is affected by hormones, age, sex and health. A hair on the head may continue to grow for several years before falling out but an eye-lash lasts for only a few months. When a hair falls out a new one grows in its place. Baldness occurs when the hair follicles cease to produce hairs.

Sebaceous Glands

Several open into each hair follicle. They produce an oily secretion called **sebum** which passes into the follicle, up to the surface, spreads over the skin and hair and helps to keep them from becoming too dry.

Hair Muscles

Each hair follicle has a muscle attached to its base. Contraction of the hair muscles due to cold or fright causes 'goose pimples' to appear.

Sweat Glands

These glands are present all over the body and secrete sweat. Each is a tube which coils at one end, then becomes a duct leading to the surface of the skin where it opens as a pore. The coiled base is surrounded by blood vessels which supply the secretory cells with water, urea and other substances so that they can produce sweat. The sweat passes along the duct and onto the surface of the skin. Its main function is to help regulate body temperature but it is also a form of excretion. 'Blackheads' occur when the pores become blocked by dirt.

Nerves and Nerve Endings

The skin is a sense organ as it is sensitive to pain, heat, cold and touch. Sensory nerves have their endings in the skin and when they are stimulated, impulses pass along the nerves to the brain where they are interpreted as pain, heat, cold, roughness, smoothness or pressure. The nerve endings may be of varying shape and a few penetrate into the epidermis. Some penetrate the hair follicle and are sensitive to movements of the hair.

Connective Tissue

This occupies the space between the other parts of the dermis and contains elastic fibres which give the skin its elasticity. These allow the skin to be stretched and then to recover its shape. The epidermis is also flexible as the furrows can be stretched out.

Fatty Tissue

Fat is stored in liquid form as droplets in the innermost cells of the dermis and accumulates more readily under some parts of the skin than others. The fat acts both as a food store and as an insulating layer.

REGULATION OF BODY TEMPERATURE BY THE SKIN

The chemical activities of the body release heat energy which is distributed around the body by the blood. At the same time heat is continuously being lost from the surface of the body by radiation, conduction and evaporation. The blood vessels and sweat glands of the skin help to keep the body temperature constant by the following means.

When the body is too hot due to vigorous activity, hot weather or illness,

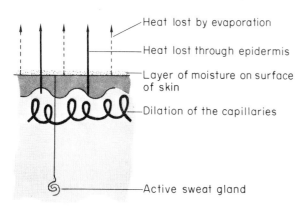

Heat lost by evaporation

Heat lost through epidermis

Layer of moisture on surface of skin

Dilation of the capillaries

Active sweat gland

1. Blood vessels in the dermis dilate allowing more blood to flow near the surface so that more heat can be lost through the epidermis by radiation and conduction. (The skin looks flushed.)

2. Sweat glands increase the rate of sweat production so that a continuous layer of moisture is produced on the skin surface. As evaporation takes place, body heat is used up.

When the body is too cold because more heat is being lost than generated,

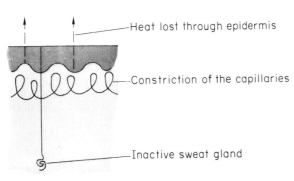

Heat lost through epidermis

Constriction of the capillaries

Inactive sweat gland

1. Blood vessels constrict allowing less blood to flow near the surface and so reducing the amount of heat lost through the epidermis by radiation and conduction. (The skin looks pale and blue.)

2. Sweat glands reduce the rate of sweat production so that little evaporation can take place, therefore little heat is lost in this way.

3. Shivering occurs, which is a reflex action. When the body temperature drops the sporadic contraction of the muscles produces heat. Goose pimples appear. In other mammals this would have the effect of raising the hair to trap a layer of air to prevent heat loss.

Chapter Eight

THE NERVOUS SYSTEM
AND SENSE ORGANS

THE NERVOUS SYSTEM

The function of the nervous system is to co-ordinate all the activities of the body in response to stimuli from both the external and internal environment.

STRUCTURE OF NERVOUS TISSUE

This tissue contains nerve cells called **neurones** (or **neurons**) which are constructed in such a way that they can relay messages called impulses from one part of the body to another. A typical neurone, for example a motor neurone (Fig. 8.1), consists of a cell body which gives rise to a number of processes that branch at their ends. One process which is longer than the rest is called the **axon** and it transmits impulses away from the cell body. The other processes are called **dendrites** and they transmit impulses towards the cell body.

Fig. 8.1 Diagram of a motor neurone. The broken lines indicate that the nerve fibre can be of great length.

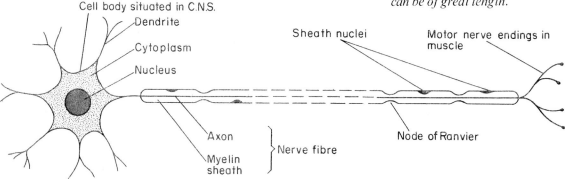

An axon may be very long and may, for example, connect the spinal cord to the big toe, and impulses only travel along the axon in one direction. It is often surrounded by a sheath of fatty substance called the **myelin sheath,** and the axon and sheath together make a **nerve fibre.** The myelin sheath is formed by the sheath cells, called **Schwann cells,** which occur in the outer region and one cell is responsible for the formation of a segment of the sheath. The junction between one segment and the next is called the **node of Ranvier.** The fatty substance the sheath contains gives the nerve fibre its white colour.

If the cell body of the **neurone** is destroyed, the cell cannot be replaced. But if the axon is cut, a new end will be able to grow out from the cut surface of the axon.

There is great variety in the shape and size of neurones but they can be classified according to their function.

1. Motor Neurones

These relay impulses from the **Central Nervous System (C.N.S.)** to muscles and glands. The cell body of a motor neurone is situated in the C.N.S. and it gives rise to a long axon (Fig. 8.1).

2. Sensory Neurones

These relay impulses from the sense organs to the C.N.S. A typical sensory neurone has the cell body situated in a **ganglion** where it gives rise to a nerve fibre which soon divides into two. The branch which goes to the C.N.S. is called the **axon,** as it relays impulses away from the cell body. The other part of the nerve fibre is called the **dendrite,** as it relays impulses towards the cell body. In Fig. 8.2 there is only one dendrite and it is longer than the axon.

Fig. 8.2 Diagram of a sensory neurone.

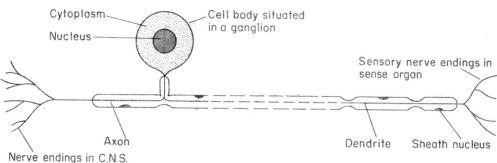

3. Connector Neurones

These relay impulses between neurones. They have relatively short axons and are found in the grey matter of the C.N.S. and in the retina (Fig. 8.9).

Nerve

A nerve is a collection of nerve fibres bound together in a bundle by connective tissue (Fig. 8.3).

Fig. 8.3 T.S. nerve.

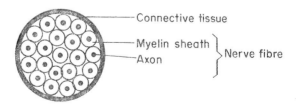

Ganglion

A ganglion is a group of nerve cell bodies situated along the course of a nerve.

Nerve Impulse

An impulse is of an electrochemical nature and passes along nerve fibres at about 120 metres per second. This is not nearly so fast as the speed of electricity (186 000 miles per sec.), and as the impulse passes along the fibre, chemical changes take place.

Synapse

The place at which two neurones meet. Although coming very close together their dendrites do not actually touch and the impulse has to cross the gap in order to pass from one neurone to the other. A neurone can use its dendrites to form synapses with one or many neurones.

The nervous system consists of:

1. Central nervous system (C.N.S.) which is the centre of co-ordination and is composed of the brain and spinal cord.

2. Peripheral nervous system composed of 43 pairs of nerves which radiate from the C.N.S. to the various parts of the body.

CENTRAL NERVOUS SYSTEM

The brain and spinal cord are very delicate parts of the body and are protected by the bony cranium and vertebral column in which they are situated. They are also enclosed in three membranes called meninges:

Dura Mater—the tough protective outer membrane.

Arachnoid—the middle layer which is a thin, net-like membrane with the spaces filled with **cerebro-spinal fluid.** This fluid helps to cushion the C.N.S. against shock.

Pia Mater—the inner membrane which contains blood vessels to supply the nervous tissue.

The C.N.S is made up of grey matter and white matter.

Grey Matter contains mainly nerve cell bodies well supplied with blood vessels.

White Matter contains nerve fibres and few blood vessels.

THE BRAIN

This is large, fills the cranium, and weighs about 1400 g (3 lb). It needs to be well protected because if the brain cells are damaged they cannot be replaced (unlike muscle and bone cells). The brain has five main parts (Fig. 8.4).

Cerebrum (fore-brain)

This forms the bulk of the brain. It is divided into two halves called **cerebral hemispheres** which are linked together at the base. The surface of each cerebral hemisphere is much folded, with an outer layer of grey matter several cells deep called the **cortex,** enclosing mainly white matter.

Because the cerebral hemispheres are relatively so much larger in man than in apes, or indeed any other animal, a human being is capable of many more complex activities such as speech, writing, memory, thinking and emotion. The cerebrum is responsible for all these and certain areas have been shown to be associated with definite activities.

Functions

1. The cortex receives impulses from the sensory organs and interprets them as smell, sight, noise, touch, warmth, cold, pain, etc.

2. Interconnection between the neurones of the cortex results in memory, reason, will-power, joy, sorrow, etc.

3. The motor neurones convey impulses for appropriate actions to be taken.

4. The cerebrum governs the activities of the rest of the brain, connector neurones linking the various parts.

Fig. 8.4 Diagram of the brain as seen from the left side. The shaded areas are those which are known to be connected with certain functions.

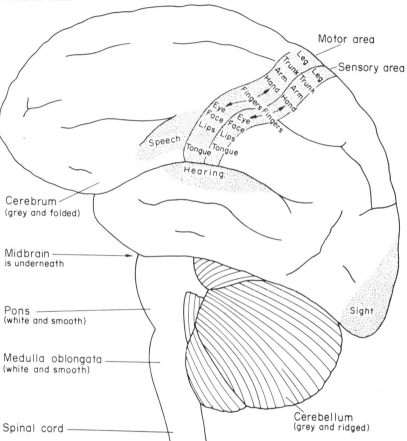

Mid-Brain

This is a small stalk, about 2 cm long and composed mainly of nerve fibres to link the cerebrum to the rest of the C.N.S.

Cerebellum

This is situated at the back of the brain, below and partly covered by the cerebrum. It is also divided into two hemispheres that are linked together. Each hemisphere has a ridged outer surface of grey matter enclosing a large mass of white matter which is so arranged that in section it appears like a tree with branches.

Function

To co-ordinate the movements of the skeletal muscles.

Pons

This region serves as a bridge linking the two halves of the cerebellum to each other and to the rest of the C.N.S.

Medulla Oblongata

This region is continuous with the spinal cord and has white matter to the outside enclosing a central mass of grey matter.

Function

It controls the automatic activity of the internal organs such as respiration, digestion, heart beat. It is also the area in which the motor fibres from the rest of the brain cross over so that those from the right side of the brain affect the left side of the body and vice versa.

The cerebellum, pons and medulla form the **hind-brain.**

SPINAL CORD

This extends from the brain to the lumbar region of the spine passing through the neural canal of each vertebra. It is cylindrical and about 45 cm long, tapering to a point. In section the spinal cord can be seen to consist of white matter surrounding a central mass of grey matter arranged in the shape of an H (Fig. 8.5).

PERIPHERAL NERVOUS SYSTEM (43 pairs of nerves)

Sensory Nerves contain nerve fibres of sensory neurones and relay impulses to the C.N.S.

Motor Nerves contain nerve fibres of motor neurones and relay impulses from the C.N.S.

Mixed Nerves contain the nerve fibres of both sensory and motor neurones.

AUTONOMIC NERVOUS SYSTEM

The autonomic nervous system is that part of the nervous system which regulates many vital functions of the body in order to maintain a stable internal environment. It consists of nerves and ganglia (the largest is the **solar plexus** situated just below the diaphragm) which link the C.N.S. with the different organs.

The autonomic nervous system functions automatically and below the level of consciousness to control such functions as heart beat, breathing, digestion, pupil size, amount of blood in the arteries, etc. It has two parts, **sympathetic** and **parasympathetic,** and they both supply nerves to the various organs, the nerves ending in either smooth muscle, cardiac muscle or glands. The two parts work in opposition, the sympathetic nerves stimulate the action and the parasympathetic nerves slow it down, e.g. sympathetic nerves speed up the rate of heart beat, the parasympathetic nerves reduce it.

HOW THE NERVOUS SYSTEM FUNCTIONS

REFLEX ACTIONS

A reflex action is an automatic and unlearned response to a

stimulus. The pathway along which this action takes place is known as the reflex arc and it involves a receptor organ to receive the stimulus, sensory (afferent) nerve fibre, C.N.S., motor (efferent) nerve fibre, and an effector organ to cause the body to take action.

The Knee-Jerk Reflex

To demonstrate this example of a reflex action a person needs to sit down with one leg crossed over the other so that the lower part of the leg hangs free. If the knee is tapped below the knee-cap the lower part of the leg will jerk forward. In this case the receptor organ is the stretch receptor which consists of dendrites of the sensory neurone being wrapped around a muscle fibre of the quadriceps femoris muscle. When the muscle fibre is stretched an impulse is caused to pass along the sensory nerve fibre to the spinal cord. Here a synapse is formed with a motor neurone, which relays the impulse to the effector organ, the quadriceps femoris. The impulse causes this muscle to contract and, as the tendon is inserted

Fig. 8.5 Diagram of the knee-jerk reflex. Arrows show the pathway of the reflex arc.

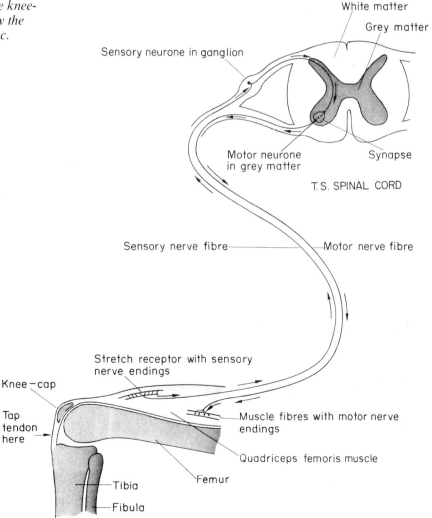

63

in the tibia, the lower leg jerks forward (Fig. 8.5). This is an automatic response over which the will has no control. In this case the brain is only aware that it has occurred after the event. However, with some other reflexes, such as the pupil reflex, the person is not conscious of the action at all.

Other examples of reflex actions are listed below.

RECEPTOR ORGAN	EFFECTOR ORGAN	ACTION
Retina, which is sensitive to light intensity.	Muscles of the iris.	Alteration of the size of the pupil; called pupil reflex.
Lining of the respiratory tract, which becomes irritated by unwanted matter.	Intercostal and abdominal muscles, which suddenly contract.	Coughing.
Skin of the foot, which contains nerve endings sensitive to pain.	Muscles of the leg.	Co-ordinated control of the muscles to withdraw the foot from a painful stimulus, such as standing on a pin.
Mouth	Salivary glands	Presence of food causes the production of saliva.

CONDITIONED REFLEX ACTIONS

When a reflex action is in response to a stimulus other than a natural one, it is called a conditioned reflex. It is an action which is acquired and not inborn, and it involves some form of learning. It can also be unlearnt or changed.

The most famous example is that of Pavlov's dogs. The dogs produced saliva at the sight or smell of food (a reflex action). Pavlov found that if a bell was rung every time before the food appeared, eventually the dogs produced saliva at the sound of the bell, whether the food appeared or not (a conditioned reflex).

It can be considered that conditioned reflexes are the basis of much human training and habit formation. Walking, talking and writing are all actions which have to be learnt before they become automatic.

VOLUNTARY ACTIONS

These are actions which are under the control of the will and require thought. They, therefore, involve the use of the grey matter of the cerebrum.

An example is shown in Fig. 8.6. A housefly is seen on some food and a decision is taken in the brain to remove it. Light rays from the food and fly enter the eye, and the retina is stimulated to send impulses to the brain. In the cerebrum these are interpreted as seeing the fly on the food. Impulses are sent from the motor area to the cerebellum to co-ordinate body movements, then to the spinal cord, which relays impulses to the muscles of the arms and legs, so that the body will be in the correct position to knock the fly off the food.

Fig. 8.6 Diagram to illustrate the voluntary action of removing a housefly from food.

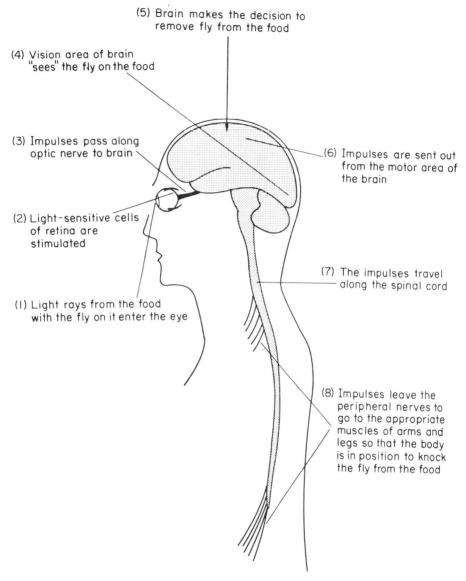

(5) Brain makes the decision to remove fly from the food

(4) Vision area of brain "sees" the fly on the food

(3) Impulses pass along optic nerve to brain

(6) Impulses are sent out from the motor area of the brain

(2) Light-sensitive cells of retina are stimulated

(7) The impulses travel along the spinal cord

(I) Light rays from the food with the fly on it enter the eye

(8) Impulses leave the peripheral nerves to go to the appropriate muscles of arms and legs so that the body is in position to knock the fly from the food

SENSE ORGANS

Sense organs are parts of the body containing sensory cells or nerve endings which respond to stimuli from the environment, each sense organ being sensitive to certain stimuli only. The eye is sensitive to light, the ear to sound, the skin to touch, pressure, pain and temperature; the nose to smell and the tongue to taste.

65

THE EYE

There are two eyes situated in bony sockets called **orbits** at the front of the skull.

The function of the eye is to receive light rays and transmit impulses to the brain where they are interpreted as 'seeing', that is, being aware of light intensity, shape, size, colour, position and movement.

STRUCTURE

The eye is spherical with a three-layered wall and the interior filled with transparent matter (Fig. 8.7). It is suspended in the socket within the orbital fat. Six muscles control movements (Fig. 8.8). When all the muscles are relaxed the eyeball faces forwards, but contraction of one or more of the muscles causes it to move so that it can receive light rays from a different direction.

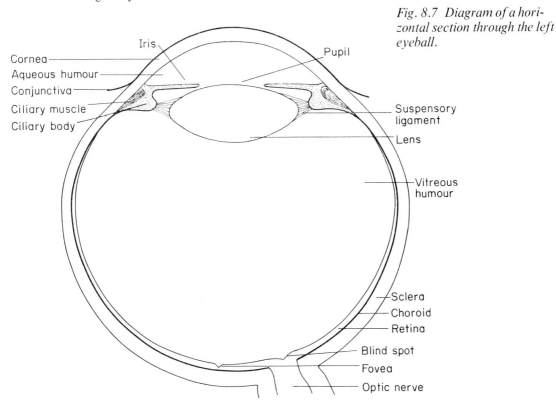

Fig. 8.7 Diagram of a horizontal section through the left eyeball.

Sclera

The tough, white opaque outer layer which protects the more delicate inner parts. It is known as the 'white of the eye'.

Cornea

The continuation of the sclera over the front part of the eye. It is transparent to allow light rays to pass inside.

Choroid

The thin middle layer containing the main arteries and veins of the eye. It also contains cells possessing a very dark pigment which prevents reflection of light within the eyeball.

Ciliary Body

A ring of thickened tissue continuous with the choroid. It contains the ciliary muscle which is concerned with the process of accommodation (p. 70) and it also produces the aqueous humour.

Fig. 8.8 Diagram of the left eye as seen from the outer side to show how it is protected and moved.

PROTECTION

MOVEMENT

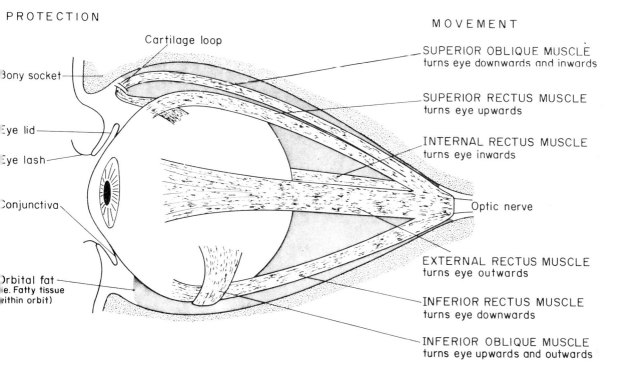

Cartilage loop

Bony socket

Eye lid

Eye lash

Conjunctiva

Orbital fat
(i.e. Fatty tissue within orbit)

SUPERIOR OBLIQUE MUSCLE
turns eye downwards and inwards

SUPERIOR RECTUS MUSCLE
turns eye upwards

INTERNAL RECTUS MUSCLE
turns eye inwards

Optic nerve

EXTERNAL RECTUS MUSCLE
turns eye outwards

INFERIOR RECTUS MUSCLE
turns eye downwards

INFERIOR OBLIQUE MUSCLE
turns eye upwards and outwards

Iris

The front part of the choroid. It is coloured blue or brown or grey etc., and contains muscle fibres, some arranged in a circular manner and others radially. The function of the iris is to regulate the amount of light entering the eye through the pupil.

Pupil

The hole in the centre of the iris through which light passes. In bright light the pupil becomes smaller because the circular muscles of the iris contract. In dim light it enlarges because the radial muscles contract.

Lens

A biconvex transparent disc which is altered in thickness and refractive power during the process of accommodation.

Suspensory Ligament

Holds the lens in position and is attached to the ciliary body.

Aqueous Humour

Transparent watery fluid in the anterior part of the eye. It is concerned with the nutrition and metabolism of the lens and cornea, which have no blood supply of their own.

Vitreous Humour

Transparent jelly which fills the posterior part of the eye. The vitreous and aqueous humours fill the eyeball and maintain its shape.

Retina

The innermost layer of the wall of the eye. It is transparent, several cells deep and contains the light-sensitive cells (Fig. 8.9).

The optic nerve enters the eyeball at the back, slightly to the inner side of the centre, and nerve fibres spread out all over the surface of the retina. Each ends in a nerve cell body (**ganglion cell**) which is linked by connector neurones to the special light-sensitive cells called **rods** and **cones** (Fig. 8.9). The cones are sensitive to colour and high levels of illumination and are more numerous in the centre of the retina opposite the pupil; the **yellow spot** located in the exact centre having only cones. The central depression in the yellow spot is

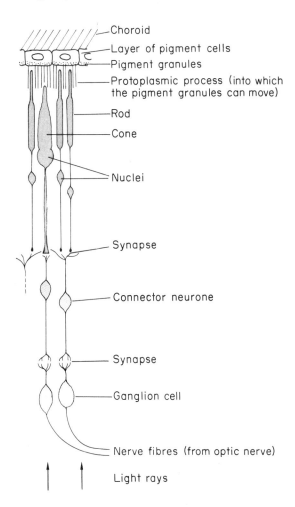

Choroid
Layer of pigment cells
Pigment granules
Protoplasmic process (into which the pigment granules can move)
Rod
Cone
Nuclei
Synapse
Connector neurone
Synapse
Ganglion cell
Nerve fibres (from optic nerve)
Light rays

Fig. 8.9 Diagram of a section through the retina.

called the **fovea.** An object is most clearly seen when light rays from it are focused directly on the yellow spot; the other objects surrounding it will be blurred. Moving outwards from the yellow spot the number of cones decreases and the rods increase. The rods are sensitive to low levels of illumination but not colour, so that in a dim light everything appears in various shades of grey. Also in a dim light the pupil is wide open which allows the light to reach the rods at the sides of the eyes. The rods contain a substance called **visual purple** which is bleached in strong light but is gradually regenerated in a dim light. Consequently, when moving from a bright light to a dim light nothing can be seen at first but as visual purple gradually redevelops it becomes possible to see. The eye is said to have become adjusted to the dark. The reverse happens when returning to a bright light. At first the eyes are dazzled but the visual purple rapidly changes and the pupil gets smaller. The eyes are now adapted to bright light.

Blind Spot

Fig. 8.10 To demonstrate the presence of the blind spot. Hold the book about 20 cm away from the face. Cover the left eye and focus the right eye on the cross. Move the book towards and away from the face. At one point the black circle will disappear. At this point the light rays from it are falling on the blind spot, and therefore the brain is not registering an image.

The point at which the optic nerve enters the eye. It is called the blind spot because there are no light-sensitive cells in this region (Fig. 8.7). Fig. 8.10 demonstrates the presence of the blind spot.

PROTECTION OF THE EYE (Fig. 8.8)

1. It lies in a bony socket called the orbit, the bone usually projects beyond the eye.

2. Fatty tissue lining the orbit forms a padding to cushion shocks.

3. The inner surface of the eyelids and the front of the eye is covered by a very thin, smooth protective membrane called the **conjunctiva.** The part covering the eyeball becomes the outer layer of the cornea. It contains numerous free nerve endings and also, if injured, has the power of rapid healing.

4. The eyelids, fringed with eyelashes, close to protect the eye from damage.

5. **Lachrymal Glands** (tear glands) situated in the upper, outer corner of the orbit continuously secrete a fluid to keep the eyeball moist. Blinking helps to distribute this fluid and any excess is drained away by a duct from each eye into the nasal cavity. Besides keeping the eye clean this fluid contains lysozyme which destroys bacteria. If the eye is irritated by foreign bodies such as dust or flies, or by chemicals such as onion juice and tear gas, the lachrymal glands secrete excessive fluid (tears) to wash the irritant out. Emotional circumstances can also be the cause of tears.

HOW THE EYE FUNCTIONS

Light rays can pass through the eyeball to the retina because the cornea, aqueous humour, lens and vitreous humour are all transparent. As the light rays pass through these substances they are refracted (bent) so that they form an image (picture) on the retina which is upside down and reversed from side to side (Fig. 8.11). The greatest refraction takes place when the light rays pass through the cornea, and the amount of refraction that takes place in the lens is varied by the lens changing shape so that the image can be focused on the retina.

Fig. 8.11 Diagram to illustrate that the eye refracts light rays to form an inverted image on the retina.

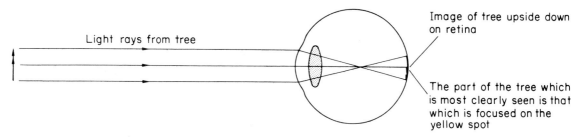

Light rays from tree

Image of tree upside down on retina

The part of the tree which is most clearly seen is that which is focused on the yellow spot

The sensory cells of the retina are stimulated by the light and impulses are sent along the nerve fibres which leave the eye in the optic nerve to go to the brain. Here they are interpreted the right way up as pictures of the objects looked at.

Accommodation of the Eye

This means the alteration of the thickness and refracting power of the lens so that light rays from a particular object can be focused on the retina. To see near objects the lens thickens and becomes more convex to cause greater refraction, i.e. the light rays are bent to a greater degree (Fig. 8.12).

Fig. 8.12 Accommodation of the eye for near objects.

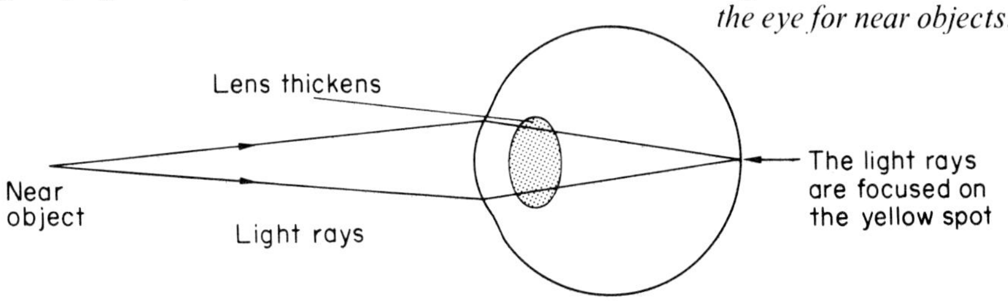

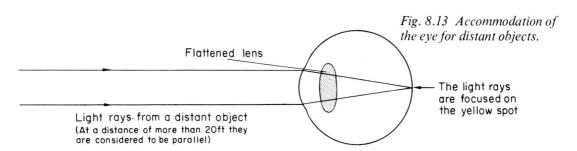

Lens thickens

Near object

Light rays

The light rays are focused on the yellow spot

Fig. 8.13 Accommodation of the eye for distant objects.

Flattened lens

Light rays from a distant object
(At a distance of more than 20ft they are considered to be parallel)

The light rays are focused on the yellow spot

To focus distant objects the lens flattens so that there is less refraction (Fig. 8.13). The shape of the lens is controlled by the

THE NERVOUS SYSTEM AND SENSE ORGANS

ciliary muscle; when this contracts the suspensory ligament slackens and the inherent elasticity of the lens causes it to become more convex. When the ciliary muscle relaxes, the suspensory ligament tightens and the lens flattens and becomes less convex. When the eye is at rest the ciliary muscle is in a relaxed state and the eye naturally focuses on distant objects.

Binocular Vision

Binocular vision is achieved by the use of both eyes together so that the separate images arising in each eye are appreciated by the brain as a single image. This achievement is not inborn, it is an acquired ability built up gradually by the child during the first few months of life.

When distant objects are viewed, there is very little difference between the images, and distance is judged by the relative size of the objects. **Stereoscopic vision** occurs when nearby objects are viewed. The two images are different and binocular vision causes the brain to appreciate a single three-dimensional image. This gives the object an appearance of depth and enables the viewer to judge distance more accurately.

EYE DEFECTS
Short Sight (Myopia)

Near objects can be seen clearly but the eye is unable to focus light rays from distant objects on the retina, so the objects appear blurred (Fig. 8.14). They are focused in the vitreous humour because the eyeball is too long (it has grown that way).

Remedy Spectacles with lenses which diverge the light rays before they enter the eye.

Fig. 8.14 Diagram to illustrate short sight and the method of correction.

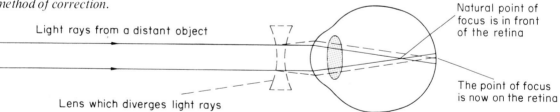

Light rays from a distant object

Lens which diverges light rays

Natural point of focus is in front of the retina

The point of focus is now on the retina

Long Sight (Hypermetropia)

Distant objects can be seen clearly but the eye is unable to focus light rays from near objects on the retina and the objects appear blurred (Fig. 8.15). The natural focus is behind the retina because the eyeball is too short.

Fig. 8.15 Diagram to illustrate long sight and the method of correction.

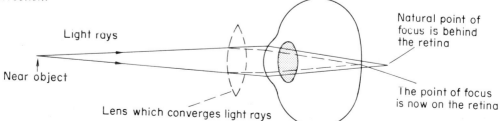

Light rays

Near object

Lens which converges light rays

Natural point of focus is behind the retina

The point of focus is now on the retina

Remedy Spectacles with lenses which converge the light rays slightly before they enter the eye.

Astigmatism

Horizontal and vertical lines are not seen with equal clarity, e.g. it is not possible to see both horizontal and vertical window bars at the same time. This is due to unequal curvature of the cornea so that light rays from different directions cannot be brought into focus at the same time, which gives a distorted image. Astigmatism can occur with long or short sight.
Remedy Spectacles with lenses thickened in the appropriate places.

Old Sight (Presbyopia)

This may occur in older people as the lens hardens and flattens. Difficulty is found in focusing nearby objects, e.g. the newspaper has to be held at arm's length in order to read it.
Remedy Spectacles with converging lenses for reading, writing and close work.

Cataract

The lens becomes cloudy, thus preventing light rays from reaching the retina.
Remedy The removal of the lens by operation and the use of spectacles afterwards to compensate.

Colour Blindness

The person is unable to see certain colours. It is very rare for a person to be totally colour-blind, usually it is a case of red and green appearing as the same colour. This characteristic is inherited, sex-linked (see p. 97), and far more common in men than women. It is probably due to a defect of cone mechanism.

THE EAR

The two ears are situated one on either side of the skull. They have two functions:
1. Hearing Sound waves are received and impulses transmitted to the brain.
2. Balance Movements affecting the balance of the body are detected and impulses transmitted to the brain.

STRUCTURE

The ear is composed of three regions (Fig. 8.16).
Outer ear—receives sound waves.
Middle ear—transmits vibrations.
Inner ear—converts vibrations into impulses which are sent to the brain. It also detects movements of the head and sends this information to the brain.

Outer Ear

A tube leading from the side of the head to the ear drum, with the opening surrounded by a flap of tissue called the **pinna** or **auricle.**

In mammals where the detection of sound is important for survival it is movable and used to collect sound waves, but in man it is relatively unimportant.

External Auditory Meatus

The name given to the tube leading down to the ear drum. It is lined by skin which gives rise to hairs for extra protection, and the glands secrete a brown wax. If wax accumulates it can cause deafness.

Tympanic Membrane (Ear Drum)

A thin sheet of tissue about 1 cm in diameter separating the outer from the middle ear.

Fig. 8.16 Diagram to show the structure of the ear.

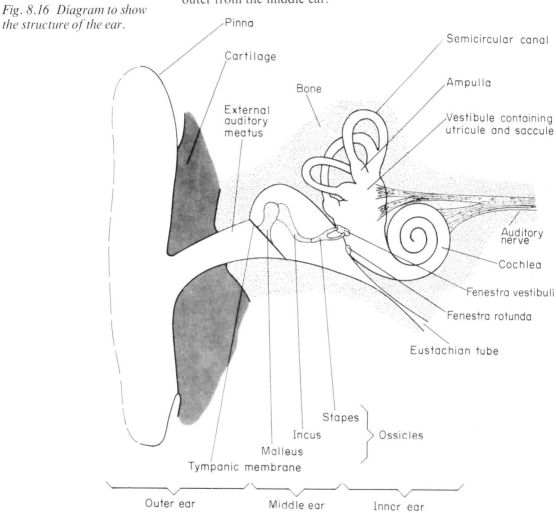

Middle Ear

An air-filled cavity surrounded by bone. It contains three small bones called **ossicles** which link the tympanic membrane with the oval window of the inner ear. Because of their shapes, these bones are known as the **malleus** (hammer), **incus** (anvil) and **stapes** (stirrup).

Eustachian Tube

The tube leading from the middle ear to the pharynx (throat). The pharyngeal end is normally closed but during swallowing it opens and allows the air pressure in the middle ear to equalise with the atmospheric pressure. If the pressures on either side of the ear drum become unequal, e.g. when landing in an aeroplane, the drum becomes stretched and causes pain. Frequent swallowing helps to equalise the pressure.

Fenestra Vestibuli (Fenestra Ovalis, The Oval Window)

An oval opening into which fits the base of the stapes. It separates the middle from the inner ear.

Inner Ear

A labyrinth of fluid-filled cavities and canals embedded in bone.

Cochlea

A spirally-coiled tube resembling a snail's shell. It is filled with fluid and is the part of the inner ear concerned with hearing. The **organ of Corti** is situated on a membrane which runs along the length of this spiral tube and contains sensory cells which are sensitive to vibrations caused by the sound waves.

Fenestra Rotunda (The Round Window)

A round opening on the side of the cochlea covered by a membrane. It allows the fluid in the cochlea to vibrate by bulging in and out in response to changes in pressure.

Semicircular Canals

These are three canals which lie at right angles to each other. Together with the **utricule** and **saccule** they form the part of the ear concerned with balance.

THE MECHANISM OF HEARING

Sound waves pass along the external auditory meatus of the outer ear and cause the tympanic membrane to vibrate. These vibrations are transmitted through the ossicles to the fluid of the cochlea via the fenestra vestibuli. Vibrations of the fluid cause the sensory cells of the organ of Corti to be stimulated to send impulses to the brain. Low notes produce vibrations which affect the sensory cells at the tip of the cochlea, whilst high notes produce vibrations which affect the sensory cells at the base. The **auditory nerve** relays the impulses to the brain where they are interpreted as hearing.

The vibrations in the fluid of the cochlea cause waves of pressure to build up. This pressure is released via the fenestra rotunda, which vibrates and passes the waves of pressure down the Eustachian tube.

THE MECHANISM OF BALANCE

The three semicircular canals are arranged in the three planes at right angles to each other and each has a swelling at one end called an **ampulla.** The ampullae, utricule and saccule all contain hair cells. These produce hairs which project into the fluid and are surrounded by a special gelatinous matter containing chalky particles called **otoliths.** When the head is moved the gelatinous matter moves within the fluid and stimulates the hair cells to send impulses to the brain. This correlates the information and if necessary sends out impulses to the skeletal muscles to restore the balance of the body. Dizziness occurs when the body has ceased to move whilst the fluid continues to do so.

THE NOSE

Besides being part of the respiratory system the nose also contains the organ of smell. This consists of special sensory cells—**olfactory cells,** which are evenly distributed amongst **supporting cells** in the membrane lining the upper region of each nasal cavity (Fig. 8.17). Each olfactory cell has a dendrite which extends to the surface to give off minute branches; the opposite end gives rise to an axon. **Olfactory glands** produce a continuous secretion to keep the surface moist.

Fig. 8.17 Diagram of a section through the membrane of the nose which contains the cells sensitive to smell.

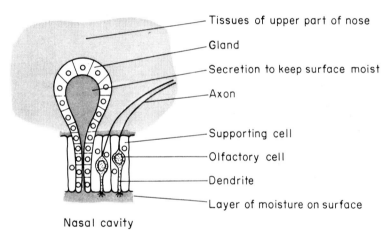

The olfactory cells are sensitive to chemicals which enter the nose in a gaseous state in the air stream. The chemicals must then go into solution in the layer of moisture before the olfactory cells can be stimulated to transmit impulses to the brain. There they are interpreted as smell.

THE TONGUE

Besides playing a part in chewing, swallowing and speech, the tongue is also the organ of taste. The upper surface is covered with projections called **papillae** of varying shapes with **taste buds** embedded in them, most commonly at the sides (Fig. 8.18). A taste bud is an oval structure with a small pore opening on to the surface. It contains 4–20 **taste cells** together with **supporting cells.** Each taste cell has a short **taste hair** which projects towards the pore.

The taste cells are sensitive to chemicals in solution and as saliva keeps the mouth moist, chemicals in the food can be dissolved. The taste cells can then be stimulated to send impulses to the brain via the nerve fibres which are linked with the taste bud. There are four tastes—sweet, sour, salt, bitter, and different areas of the tongue are particularly sensitive to different tastes. Sweet tastes are most easily detected at the tip of the tongue, bitter at the back, sour at the edge and salt on the tip and the edge.

Fig. 8.18 Left: *Section through the upper surface of the tongue.* Right: *Section through a taste bud.*

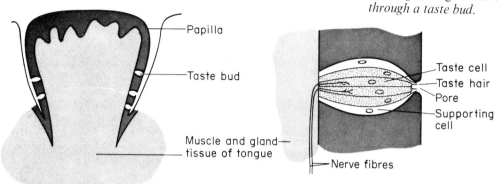

Flavour is a combination of taste and smell. When a person is unable to smell, for example when suffering from a cold, food appears to have little flavour.

THE SKIN

The skin is a sense organ as it contains nerve endings sensitive to touch, pain, warmth and cold, by which the body is made aware of changes in the external environment.

There are many different kinds of nerve endings in the skin (Fig. 7.4). Some branch in the epidermis, others end in large swellings in the inner dermis, or are associated with hair follicles so that the body is made aware of any movement of the hair. Other nerve endings take different forms.

There is no conclusive evidence that the different types of nerve endings respond to different stimuli, i.e. that one type is sensitive to warmth, another to pain, etc.

Chapter Nine

CHEMICAL CO-ORDINATION

An **endocrine gland** is often called a **ductless gland** as it has no duct and its secretions pass directly into the blood stream for distribution around the body.

A **hormone** is a chemical substance which is produced in one part of the body and transported in the blood stream to affect tissues or organs in another part of the body.

Fig. 9.1 Endocrine glands in the male.

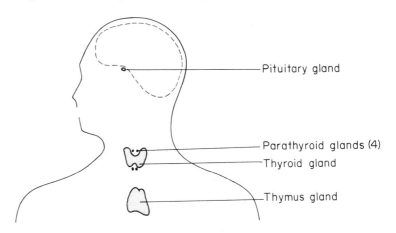

Pituitary gland

Parathyroid glands (4)

Thyroid gland

Thymus gland

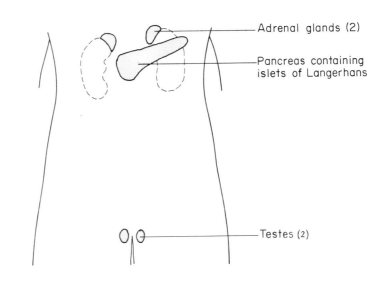

Adrenal glands (2)

Pancreas containing islets of Langerhans

Testes (2)

Endocrine glands secrete hormones which act as chemical messengers controlling and co-ordinating such processes in the body as metabolism, growth and reproduction. Different endocrine glands secrete different hormones and although each hormone travels all over the body in the blood stream, it will only affect particular cells and in a certain way. The main endocrine glands are the pituitary, thyroid, parathyroids, adrenals, islets of Langerhans found in the pancreas, and the testes or ovaries (Fig. 9.1).

THE PITUITARY GLAND

The pituitary gland is a small gland about the size of a pea attached by a stalk to the base of the brain. This is the so-called 'master' gland, as its hormones regulate the activity of the other endocrine glands and influence all the metabolic processes. It secretes a large number of hormones including those which:

Affect the rate of growth, under-production resulting in a dwarf and over-production in a giant (the **growth hormone**).

Affect the development and functioning of the sex organs and secondary sexual characteristics.

Influence blood pressure.

Influence urine production.

THE THYROID GLAND

The thyroid gland is bi-lobed and situated in the neck just below the larynx, surrounding the front part of the trachea. It contains a large amount of iodine which is needed to produce the hormone **thyroxine.** This hormone controls the metabolic rate.

A child whose thyroid gland does not function develops into a mentally retarded dwarf called a **cretin.** If the condition is diagnosed at an early age, a cure is possible by giving the child thyroid extract obtained from sheep.

An adult whose thyroid gland ceases to function suffers from a slowing-up of all the bodily processes called **myxoedema.** This too can be cured by taking regular doses of thyroid extract.

If the thyroid gland is over-active and produces too much hormone all the metabolic processes speed up causing the person to become excitable and apprehensive. The cure is often to remove part of the thyroid gland.

Another condition called **goitre** may occur if the diet is deficient in iodine. This causes the thyroid to become extremely swollen and bulge out on the front of the neck. Goitre occurs particularly in people living in mountainous regions where the drinking water is deficient in iodine. In England it used to be frequent in the Peak district and was known as Derbyshire Neck. To remedy this condition table salt sold there is iodized, that is, a small amount of iodine in the form of potassium iodide is added to it.

THE PARATHYROID GLANDS

The parathyroid glands lie in the neck near the thyroid. Usually there are four small glands but their number, size and position varies. They produce the **parathyroid hormone** which regulates the calcium content of the blood (p. 80).

ADRENAL GLANDS (SUPRARENAL GLANDS)

The two glands are situated above the kidneys. Each has two parts—an outer cortex surrounding a central medulla.

The **adrenal medulla** secretes the hormone **adrenaline.** In times of stress (e.g. fear, anger) large quantities of adrenaline are quickly released into the blood stream to prepare the body for action. It does this by increasing the rate of heartbeat, converting glycogen to glucose, and diverting blood from skin to muscles (the person turns pale).

The **adrenal cortex** produces several hormones, one being **cortisone,** and their function is to keep the body in a state of chemical equilibrium.

ISLETS OF LANGERHANS

These are small groups of cells scattered amongst the tissue of the pancreas which secrete the hormone **insulin.** This hormone is important in helping to control the amount of sugar in the body. It causes excess blood sugar to be converted into glycogen in the liver or muscles, for storage until required.

If the islets of Langerhans fail to secrete sufficient insulin, the person will suffer from **diabetes,** some of the symptoms being weakness, loss of weight, loss of sugar in the urine and ultimately death. Fortunately, diabetes can now be kept under control by regular injections of insulin.

TESTES AND OVARIES

Besides producing the sex cells they are also endocrine glands secreting sex hormones. The testes secrete the male hormones called **androgens** which control the development of the male sex organs and secondary sexual characteristics. The ovaries secrete the female hormones called **oestrogens** which control the development and functioning of the female sex organs and secondary sexual characteristics, and also **progesterone** which controls pregnancy.

THE THYMUS GLAND

This is no longer regarded as an endocrine gland, but it has other functions (p. 32).

ACTIVITY OF ENDOCRINE GLANDS

Activity of endocrine glands is controlled by:

1. Hormones from one gland influencing the activity of another gland, e.g.
    (a) One of the pituitary hormones influences the production of the thyroid hormone, and this in turn influences the activity of the pituitary gland.
    (b) The pituitary gland produces **'egg-production' hormone** which stimulates the ovary to produce an egg. After ovulation the ovary produces **'egg-suppressor' hormone** which prevents the pituitary from secreting egg-production hormone. As the supply of egg-suppressor hormone diminishes, the supply of egg-production hormone increases, and about 28 days later, another egg is produced.

2. Other chemicals, e.g.

The parathyroid hormone maintains the amount of calcium in the blood at a constant level. A fall in the amount of blood calcium stimulates the parathyroids to produce more hormone to increase the supply of calcium in the blood.

3. Nervous system, e.g.

Nervous stimulation causes the adrenals to release adrenalin into the blood.

INTEGRATION

The body can only function properly if all its parts work in unison. Communication between the organs, tissues and cells is by means of the nervous system and by chemical messengers. There is close co-operation between the nervous system and the endocrine system and they are in a constant state of interaction.

The Nervous System

This monitors both the external and internal environments. The central nervous system correlates the information and the appropriate nerves relay messages rapidly through the body by means of impulses. Response to the impulses is by muscles or glands. Generally, the nervous system is used to make rapid internal adjustments and to control a particular activity over a short period of time.

The Endocrine System

Hormones or other chemicals are generally used to regulate continuous or long-term processes. These chemical messengers travel in the blood to all parts of the body and, unlike impulses, not directly to the part they affect. Response to chemical messages is by glands or by the nervous system.

HOMEOSTASIS

Homeostasis is the maintenance of a steady state.

Despite continuous changes in the external environment, the internal environment, i.e. the environment within the body, needs to be kept constant. Each cell attempts to maintain a steady state, and so does the body, in order that the cells and the body can function under optimum conditions. To achieve this, various activities are self-regulating and operate by means of **feed-back mechanisms.**

Examples of Feed-back Mechanisms in the Body

1. Respiration

Respiration is controlled by the **respiratory centre** in the brain which sends impulses alternately to the muscles controlling inspiration and to those controlling expiration. This ensures regular breathing.

When the body works harder more oxygen is required, more carbon dioxide is produced, and a feed-back mechanism comes into action to increase the rate and depth of breathing. The increased concentration of carbon dioxide in the blood passing through the brain causes the respiratory centre to speed up the rate and intensity of its impulses to the muscles of the thorax and diaphragm. The result is faster, deeper breathing, which increases the oxygen supply and eliminates more carbon dioxide. Breathing returns to normal as the amount of carbon dioxide in the blood falls.

2. *Temperature*

The control centre for body temperature is in the brain and it is sensitive to the temperature of the blood. If the blood reaching the brain is cooler or warmer than normal, impulses will be sent to the skin to take appropriate action (see p. 57).

3. *Production of Eggs* (see p. 79).

4. *Blood Calcium* (see p. 80).

Chapter Ten

REPRODUCTION

Reproduction is the process by which new individuals are produced.

Man, like all vertebrates, reproduces sexually. There are two sexes, male and female, with differing reproductive organs, each producing special reproductive cells. The male produces spermatozoa in the testes and the female produces ova in the ovaries.

THE MALE REPRODUCTIVE SYSTEM

The male reproductive system is designed to make sperm and deposit them in the female's body. It is situated at the lower end of the abdomen and consists of two testes, tubes (ducts) along which the spermatozoa pass, glands and penis (Fig. 10.1). The two testes develop inside the abdomen of a male foetus but by the time the baby is born they have moved down into an extension of the abdominal cavity called the **scrotum.**

Testis

An oval body about 5 cm long. It is divided up into a large number of compartments each containing 1–4 minute coiled tubules in which the spermatozoa are produced. The testis also produces hormones (see p. 79).

Epididymis

A much coiled tube for storing the sperm which collect there from the tubules. Whilst in the epididymis the sperm are inactive.

Vas Deferens

A white muscular tube which leads from the epididymis up into the abdominal cavity where it joins the urethra just below the bladder.

Seminal Vesicle

A gland which opens into the vas deferens just before it joins the urethra. It produces a thick secretion to dilute the spermatozoa and to provide them with food for energy when they become motile.

Prostate Gland

This gland surrounds the junction where the vas deferens from each testis joins the urethra. It secretes a fluid which activates the spermatozoa.

82

Penis

The organ of copulation, that is, the organ which deposits spermatozoa in the female's body. The urethra passes through it and serves as a duct for both urine and semen. The penis is normally small and soft but before copulation can take place it becomes filled with blood which makes it hard and erect. This enables it to penetrate the vagina of the female to deposit spermatozoa as near as possible to the uterus.

Fig. 10.1 Diagrams of the male reproductive system; (a) front view, (b) side view. Arrows indicate the path of the spermatozoa.

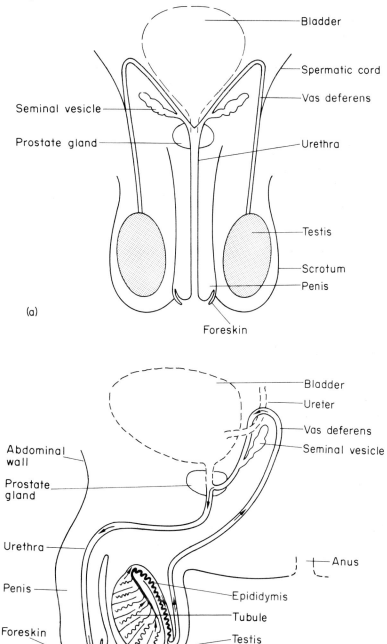

Semen

The semen is the fluid which passes from male to female. During copulation spermatozoa pass from the epididymis along the vas deferens to be mixed with fluid from the seminal vesicles and prostate gland before being ejaculated as semen into the vagina of the female.

THE FEMALE REPRODUCTIVE SYSTEM

The female reproductive system is designed to make ova, receive sperm, accommodate and feed and protect the developing baby, and to give birth. It is situated in the lower part of the abdominal cavity and consists of two ovaries, two Fallopian tubes, uterus and vagina (Fig. 10.2).

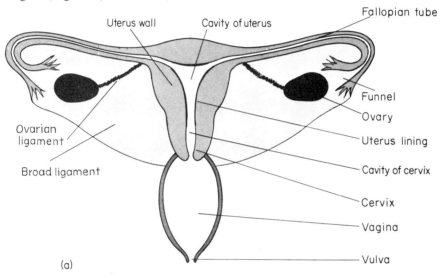

(a)

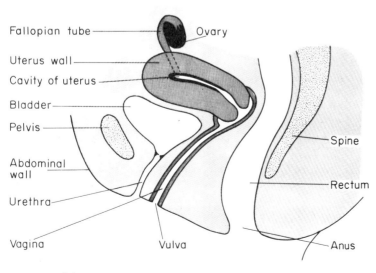

(b)

Fig. 10.2 Diagrams of the female reproductive system; (a) front view, (b) median longitudinal section through the lower abdomen to show the position of the female reproductive organs. Note the narrow cavity of the uterus.

Ovary

An almond-shaped body about 3 cm long held in place by liga-ments. The function of the ovary is to produce ova and after puberty one ovum (rarely two or more) is released from the ovaries every month during the female's reproductive life of about thirty years. Each ovum develops in a special group of cells called a **Graafian follicle.** When fully developed the follicle bulges out on the surface of the ovary, bursts, and releases the ovum.

Ovulation

The release of the ovum from the ovary. After ovulation the follicle wall thickens and the follicle is known as the **corpus luteum** or **yellow body** which produces the hormone **progesterone** to prepare the uterus to receive and implant the fertilised ovum. If the ovum is not fertilised the corpus luteum degenerates. If fertilisation takes place the corpus luteum persists during pregnancy and continues to secrete progesterone. In a section through an ovary it is possible to see the various stages of development of different follicles (Fig. 10.3).

Fig. 10.3 Diagrammatic section through the ovary.

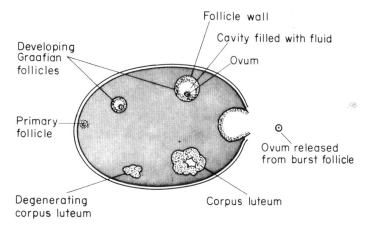

Fallopian Tube (oviduct)

A muscular tube about 12 cm long which opens into the uterus. The other end of this tube is funnel-shaped and fringed and at the time of ovulation the funnel is in close contact with the surface of the ovary. The ovum is moved along the tube by peristalsis and by the ciliated lining.

Uterus (womb)

A pear-shaped organ with a thick muscular wall and a central flattened cavity. It is well supplied with blood vessels and the cavity is lined with a thick soft mucous membrane. During pregnancy the muscular wall can enlarge to an enormous extent. The lower narrower part of the uterus is called the **cervix** (neck of the womb) which opens into the vagina.

Vagina

The passage which leads from the uterus to the exterior; the opening is called the **vulva.**

MENSTRUATION

Periodically there is a discharge from the body of blood together with fragments of the uterus lining. It begins during puberty at the age of about 13–15 years and ends with the menopause usually between the ages of 45 and 50. Menstruation is suspended during pregnancy and for a few months afterwards.

Menstrual Cycle

A regular sequence of events in the ovaries and uterus which ensures that the uterus lining is ready to receive a fertilised ovum. It is controlled by hormones from the ovaries and pituitary gland and takes about 28 days. An approximate time-table is given, but it should be remembered that this can vary with different females and also from month to month in the same female.

DAYS
- 1–5 Bleeding occurs which removes the uterus lining.
- 6–12 A Graafian follicle matures and secretes hormones called oestrogens which cause the uterus lining to be renewed.
- 13–15 Ovulation.
- 16–28 The corpus luteum develops and secretes the hormone progesterone which prepares the uterus lining for the reception of a fertilised ovum. If the ovum is not fertilised the lining of the uterus wall degenerates. If the ovum is fertilised it becomes implanted in the uterus wall.

COPULATION

Copulation is the term applied to the physical act of mating of a male and female mammal. Sexual intercourse is a term usually applied only to humans. In most mammals mating only occurs during a certain stage of the menstrual cycle. Sexual intercourse between a man and a woman, however, may occur at any time, and it is generally an emotional event, involving such feelings as love and consideration for the other partner.

FERTILISATION (CONCEPTION)

The fusion of ovum and spermatozoon is called fertilisation (Fig. 10.4). This takes place in a Fallopian tube within 12 hours of sexual intercourse and within 8 hours of ovulation. During intercourse about 300 million spermatozoa are deposited in the female's vagina. Each spermatozoon is a minute body consisting of a head containing a nucleus and a long whip-like tail. It is now capable of independent movement and moves around at random. Should it manage to penetrate the uterus and a Fallopian tube and contact an ovum then fertilisation can occur. Large numbers of spermatozoa need to be produced so that one of them may by chance contact the ovum.

Fig. 10.4 Left: *a spermatozoon;* Right: *fertilisation of an ovum by a spermatozoon.*

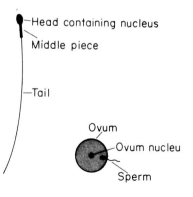

Head containing nucleus

Middle piece

Tail

Ovum
Ovum nucleu
Sperm

PREGNANCY

This follows fertilisation and involves development of the fertilised ovum accompanied by great changes in the body of the mother.

DEVELOPMENT

After fertilisation in the Fallopian tube the ovum begins to divide, first into two cells, then into four, etc., and by the time it reaches the uterus a few days later it will be a spherical mass of cells. By now the uterus wall is ready to receive it and it burrows into the thickened lining. This process is called **implantation** and once safely in place the ovum is supplied with food and oxygen from the uterus wall.

Fig. 10.5 Left: *diagram of an embryo a few weeks old;* Right: *diagram of a foetus just before birth.*

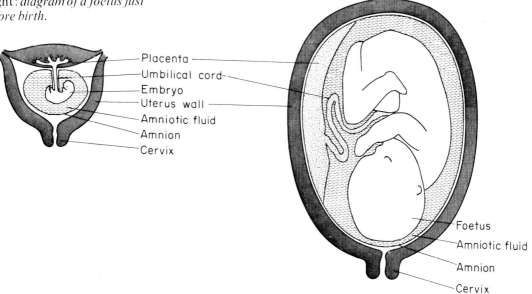

Placenta
Umbilical cord
Embryo
Uterus wall
Amniotic fluid
Amnion
Cervix

Foetus
Amniotic fluid
Amnion
Cervix

The ovum now grows rapidly with the central cells developing into the embryo whilst the remainder are concerned with the development of the placenta, umbilical cord and protective membranes (Fig. 10.5). The **placenta** grows over and penetrates into the lining of the uterus wall to become the organ of transference of materials between mother and child. It also anchors the embryo in the uterus and when fully developed it is about 12 cm in diameter and 2·5 cm thick in the centre. The umbilical cord links the abdominal region of the embryo with the placenta and grows to be about 50 cm long and 2 cm in diameter. Blood vessels from the embryo lead through the umbilical cord to branch in the placenta and become surrounded by the mother's blood. Although the blood of the mother and embryo are now very close together they do not actually meet, being separated by three very thin membranes. Exchange of materials between mother and embryo can take place by diffusion across the membranes. Oxygen, amino-acids, glucose, etc., diffuse from mother to embryo, carbon dioxide and other waste products diffuse in the opposite direction. The placenta is in fact the organ by which the embryo performs the functions of respiration, nutrition and excretion. At an early stage of development the embryo becomes surrounded by protective membranes, the most important of these is called the **amnion.** The membranes form a sac which becomes filled with **amniotic fluid** and this helps to cushion the embryo against shocks.

After a few weeks the embryo begins to show a human form and is now called the **foetus.** It is about 2·5 cm (1 inch) long at this stage and it continues to grow and develop so that about nine months after conception (fertilisation) a fully-formed baby is ready to be born (Fig. 10.5). The uterus has also had to enlarge to accommodate the growing foetus. During the second half of pregnancy the heart beat can be detected and movements of the foetus within the uterus can be felt.

BIRTH

Before birth the baby gets itself into the correct position with its head downwards in the uterus. As birth begins the cervix gradually widens and the baby is slowly pushed through by the rhythmical contractions of the uterus wall. When the contractions occur the mother feels pains called **labour pains.** At first these occur about every 20–30 minutes but the time between the contractions shortens as they become stronger. At some time during this stage the amnion bursts releasing the amniotic fluid, often referred to as the 'breaking of the waters'.

When the cervix has opened contractions by the uterus push the baby head first through both cervix and vagina, the mother helping to push by using her diaphragm and abdominal muscles.

As soon as the baby emerges air rushes into the lungs, fills them up, breathing begins, and the first thing the baby usually does is to cry. It is still connected to the mother by the umbilical cord but before this can be cut, two ligatures must be tied. The cord can then be cut between them about 5 cm from the baby.

Contractions of the uterus now start again to push out the placenta, called the **afterbirth.**

A few days after birth the piece of umbilical cord attached to the baby shrivels and falls off, leaving the **navel.** It takes about a month for the uterus and abdominal muscles of the mother to return to their normal size.

LACTATION

The mother produces milk in order to feed the baby. This is produced in the **mammary glands** (breasts) and sucked out through the nipple by the baby. The breast consists of 12–20 compartments each containing a system of branching tubes and with a duct which opens onto the surface of the nipple (Fig. 10.6).

During pregnancy the breasts enlarge but the glands do not secrete a large quantity of milk until about 3 days after birth. Before this they produce a small amount of watery fluid called **colostrum** which has a high protein and low fat and sugar content. Breast feeding can continue for several months and the milk is a complete food for the child for the first few weeks of life.

ADVANTAGES OF BREAST FEEDING

Breast milk is the natural food for a baby. It contains all the essential ingredients (carbohydrates, fat, protein, vitamins and minerals) in the right proportions and, as the baby grows, the proportions of the various ingredients change to meet the changing needs of the baby.

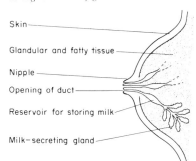

Fig. 10.6 Vertical section through a mammary gland.

Skin

Glandular and fatty tissue

Nipple

Opening of duct

Reservoir for storing milk

Milk–secreting gland

Cow's milk differs from human milk in having different types of fat and protein, less sugar, and more protein, sodium and other minerals. Powdered baby-milk is made from cow's milk which has had its composition altered during manufacture to make it more suitable for babies.

1. Breast-fed babies receive **milk of the correct composition,** whereas bottle-fed babies may receive milk which is not right for their needs. When using milk powder, there is a tendency to think that it is 'good for the baby' to add more powder than the instructions say. The 'strong' feed contains too much sodium which makes the baby cry with thirst, and he may be given another feed. Not only will he get too fat, but a high sodium level in a baby can be very dangerous.

3. Breast-fed babies have **less gastroenteritis** than bottle-fed babies. When bottle-feeding it is necessary to sterilise the bottle, teats and milk.

4. Human milk contains antibodies and other factors which help to **protect** the baby against disease.

5. Lastly, and most important, a mother who breast feeds spends a long time each day in very close and warm contact with the baby. This gives time for a **bond of affection** to develop between mother and baby which is most important for the future well-being of the child. Care needs to be taken to give bottle-fed babies the same attention.

Weaning

The gradual changeover in diet from milk to mainly solid food. This is usually completed by the time the child is nine months old.

CHILDHOOD

The stage during which the sex organs remain immature.

PUBERTY

The stage during which the sex organs mature and become capable of producing either spermatozoa or ova. Puberty begins at about 13–14 years of age in boys and 11–12 in girls and is accompanied by the development of secondary sexual characteristics.

In the male: shoulders broaden, the voice 'breaks', hair grows on the face, chest, armpits and pubic region.

In the female: pelvis widens, breasts develop, hair grows in armpits and pubic region.

SPECIAL NEEDS OF THE PREGNANT MOTHER

1. Diet A balanced diet containing a high proportion of proteins, vitamins and minerals is necessary to feed the mother and to provide food for growth and development of the foetus. An insufficient supply of these substances may cause the baby to obtain them from the tissues of the mother, thus weakening the mother. On the other hand if the mother eats too much carbohydrate and fat she may become overweight, which will make her unhealthy.

2. Clothing Comfortable clothing should be worn. If clothes are too tight they may restrict the blood circulation or breathing. Low heeled shoes help the mother to keep her balance and thus reduce the risk of her falling over.

3. Exercise Gentle daily exercise is necessary to keep the mother healthy, e.g. walking, housework. Special exercises may be done to strengthen the abdominal muscles and to improve breathing, both of which can be useful during labour.

4. Rest As the extra weight that the mother has to carry increases so does the necessity for a period of daily rest with the feet up. This helps to prevent backache, varicose veins and undue tiredness.

5. Antenatal care This means looking after the health of the mother and checking the progress of the developing child before it is born. Antenatal clinics have been set up for this purpose so that any abnormality can be detected which may interfere with the unborn child or with childbirth. The mother's blood is tested for blood group, rhesus factor and anaemia, and a routine check is kept on blood pressure. The urine is tested for sugar to detect any case of diabetes. It is also tested for protein (albumin) as this may indicate disease.

BIRTH CONTROL

The world's population is increasing at a very rapid rate—at the present rate it will double in the next 35 years—due to lower infant mortality and longer life expectancy. Because of the threat to general living standards, many authorities consider it essential to encourage birth control (contraception).

Pregnancy follows when a sperm fertilises an ovum in the Fallopian tube and the ovum becomes implanted in the uterus. The various methods of birth control are all aimed at preventing part, or all, of this sequence of events from happening.

1. Abstention from intercourse.

2. Male sterilisation—a very simple operation in which the vas deferens are cut and tied to prevent sperm from joining the semen.

3. Female sterilisation—the Fallopian tubes are cut and tied to prevent the ovum from reaching the sperm.

4. The condom (sheath, french letter) fits over the penis to prevent the entry of sperm into the vagina.

5. The cap (diaphragm) fits over the cervix to prevent entry of sperm into the uterus.

6. The coil (loop, intra-uterine device, I.U.D.) A plastic or metal coil or loop is inserted into the uterus. This works, but it is not known why.

7. The pill prevents the release of the ovum from the ovary by altering the balance of the hormones oestrogen and progesterone.

8. Rhythm method (safe period) Intercourse is restricted to those days of the menstrual cycle in which conception is less likely to take place.

9. Withdrawal method (coitus interruptus) The penis is withdrawn before semen is ejaculated. Not a very reliable method, because semen can leak from an erect penis before ejaculation takes place.

N.B. Family planning clinics give advice and information on birth control.

Chapter Eleven

INHERITANCE

Although each child is an individual he will possess certain characteristics similar to those of his parents, brothers and sisters, grandparents and other 'blood relations'. He is said to have inherited these characteristics and will have obtained them from the sex cells of his parents.

Inheritance is the transmission of characteristics from parents to offspring, and applies to both plants and animals.

Genetics is the study of inheritance. The characteristics are transmitted by genes contained on the chromosomes.

Chromosomes are thread-like structures that occur in pairs in the nucleus of the cell. The two chromosomes that make up a pair are identical and every cell of the body (except the gametes) will have exactly the same number of chromosomes, e.g. man has 23 pairs, or, to put it another way, 2 sets of 23 chromosomes which have paired off (one set originated from the mother, the other from the father).

Genes occur along the length of the chromosome and contain DNA (deoxyribonucleic acid). DNA is the hereditary material. It controls the structure and functioning of the cells of the body and hence of the body as a whole. It appears to be able to do this by producing a substance called RNA (ribonucleic acid) which leaves the nucleus and passes into the cytoplasm where it influences the production of enzymes. These control the chemical changes which take place in the cell and determine the size, shape and structure of the cell and the way it functions, i.e. its characteristics. The characteristics of an individual depend upon the characteristics of its cells.

The chromosomes occur in identical pairs, therefore the genes will occur in pairs, both genes of a pair affecting the same characteristic.

Meiosis (Fig. 11.1) is the process of cell division in which gametes with only one set of chromosomes are produced. Meiosis occurs in the testes to produce spermatozoa (sperm) and in the ovaries to produce ova (eggs).

Gametes are sex cells, the spermatozoa and the ova, but as each sex cell only has one set of chromosomes it cannot develop on its own. Therefore sexual reproduction must take place and a spermatozoon must fertilise an ovum, thus restoring the correct chromosome number, before a new individual can develop.

Zygote—the result of fertilisation. The two gametes have united to form one cell with two sets of chromosomes which can now give rise to a new individual.

Mutation—a change in the genetic make-up resulting in the development of new characteristics.

Gene mutations   Generally a gene is a very stable structure and during cell division it makes an exact replica of itself. Occasionally, however, a mistake occurs and the structure of the gene is altered. This produces a new characteristic which may be beneficial or harmful to the organism. If a mutation occurs in a body cell the variation will appear in all the cells which descend from it. If the mutation occurs in a reproductive cell it may be transmitted to offspring to produce an individual with a characteristic totally unlike any of those of either parent.

Chromosome mutations   These occur if the chromosomes fail to divide properly during cell division with the result that the new cells have more or less chromosomes than is normal. **Mongolism** in humans is the result of having 47 chromosomes instead of the normal 46.

Fig. 11.1 Meiosis

Meiosis is also called **reduction division,** *as the number of chromosomes is reduced by half during the process. The diagrams illustrate chromosome behaviour during meiosis. Only two pairs of chromosomes are shown. The chromosomes which originated from one parent are shown as thick lines, those from the other parent as thin lines.*

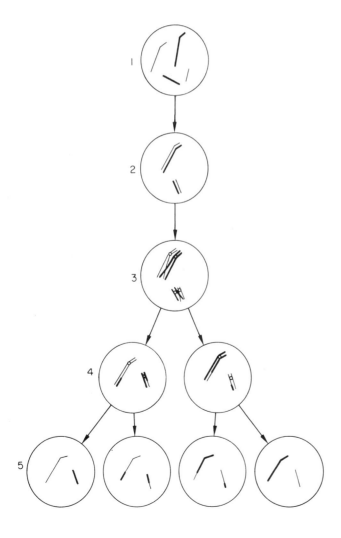

1. The chromosomes appear as single threads.

2. Identical chromosomes pair off.

3. Each chromosome divides lengthways into two **chromatids.** *The chromosomes coil around each other and parts of the chromatids are exchanged. This process is called* **crossing-over** *and it is important because genetic material is exchanged between chromosomes, the result being new combinations of genes.*

4. The chromosomes now separate, with one member of each pair going into a new cell. These two cells only contain half the number of chromo-somes possessed by the original cell.

5. The cells divide again. In this division the chromatids separate and the cells which are produced are the gametes. Because of crossing-over and the random way in which the pairs of chromosomes separate into the daughter cells in stage 4, a gamete contains genes from both parents.

GENETICAL TERMS

Phenotype	Appearance of the individual.
Genotype	Genetic constitution of the individual.
Homozygous	Possessing a pair of identical genes.
Heterozygous	Possessing a pair of dissimilar genes.
Alleles (Allelomorphs)	Different forms of the same gene which occupy the same relative position on similar chromosomes.
Dominant	A dominant gene is one which produces the same effect when present in a single dose (heterozygous) as it does in a double dose (homozygous).
Recessive	A recessive gene has no effect on phenotype unless homozygous.
Pure-Bred	Offspring having the same phenotype and genotype as the parents.
Hybrid	An individual resulting from a cross between parents that are genetically unalike.

GENETIC SYMBOLS

P = parents F_1 = offspring of parents
x = mating F_2 = offspring of F_1
G = gametes

Capital letters are used to represent dominant genes, e.g. T = tallness.
Small letters are used to represent recessive genes, e.g. t = dwarfness.

MENDEL

Gregor Mendel lived in the 19th century and was an Augustine monk in Brunn (a town in Czechoslovakia now called Brno). Mendel noticed how regularly certain characteristics occurred in particular species of plants and animals. He concluded that these characteristics must be passed, in some way, from one generation to another, and decided to investigate how this occurred. His experiments have formed the basis of all modern work on heredity. He published his results in 1866, but it was not until 1900 that scientists began to appreciate the significance of his work.

Mendel chose for his experiments pea plants, which are capable of self-fertilisation. He discovered that if he bred several generations of pea plants in this way, they began to breed true.

He found that he could breed both tall and dwarf plants. Using large numbers of plants he allowed these types to breed true for two generations and then cross-pollinated them. The resulting offspring were all tall. However, he then found that if he allowed this F_1 generation to self-pollinate, the resulting offspring were a mixture of tall and dwarf plants in the ratio of 3:1.

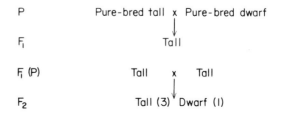

93

These results can be explained if several assumptions are made.

1. There is a gene for tallness and a separate gene for dwarfness.

2. Tallness is dominant to dwarfness. It is convenient to represent the dominant gene for tallness as T, and the recessive gene for dwarfness as t.

3. Parental chromosomes have two genes which are split up at meiosis when the gametes are formed. Thus a true-breeding tall plant we can represent as TT and a true-breeding dwarf plant will be tt. The gametes of the tall plant will each contain one gene for tallness, T, and similarly the gametes of the dwarf plant will each contain one gene for dwarfness, t.

When the parents are crossed the gametes will separate and recombine.

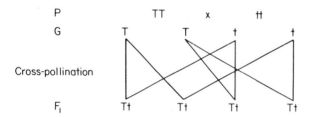

All the F_1 generation will therefore be tall.

When members of the F_1 generation are self-pollinated the gametes will separate and recombine again.

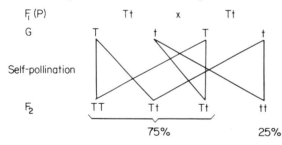

The ratio of tall to dwarf plants will therefore be 3:1.

Mendel himself was not able to draw all these conclusions, as he did not have the benefit of present day techniques and knowledge. However, from his experiments he did draw one firm conclusion, now known as **Mendel's first law of inheritance.** This states that of a pair of contrasted characteristics, each character separates out in the gametes and can recombine with either of another pair.

HYBRIDS

In the case of Mendel's peas the F_1 generation were hybrids, as, although they were tall plants, they also had a gene for dwarfness. Thus they were heterozygous for this characteristic.

If hybrids are self-fertilised or crossed with another hybrid of the same genotype, the resulting offspring are a mixture of 75% tall and 25% dwarf, as shown in the previous section.

Hybrids can also be crossed with dwarf plants. If the hybrids have been originally produced from a cross between pure-bred tall and dwarf plants, crossing them with dwarf plants is often referred to as a **back-cross**.

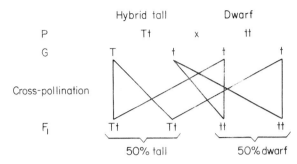

This type of back-cross can be used to check the genotype of a suspected hybrid, as a hybrid (heterozygous) tall plant is no different externally from a pure-bred (homozygous) tall plant. If the offspring of a cross between a genetically unknown tall plant and a pure-bred dwarf plant are 50% tall and 50% dwarf then this shows that the tall parent is a hybrid.

If hybrids are crossed with homozygous tall plants, the resulting offspring will all be tall.

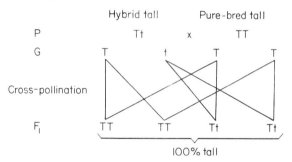

However half of these will be heterozygous and half homozygous, and the only way to distinguish between these two types will be to back-cross them with dwarf plants.

HUMAN GENETICS

Most human characteristics are due to more than one pair of genes, and this makes them much more difficult to study. However a few characteristics are due to a single pair of genes, e.g. the ability to taste phenylthiocarbamide (PTC).

To some people it tastes very bitter and to others it has no taste. Tasting is dominant to non-tasting, and therefore tasters can be homozygous or heterozygous, whilst non-tasters are always homozygous.

If we represent tasting by T and non-tasting by t, we can work out the theoretical results of various possible marriages.

1. If two homozygous tasters marry, the children will all be homozygous tasters.

2. If two non-tasters marry, the children will all be non-tasters.

3. If a homozygous taster marries a non-taster, the children will all be tasters.

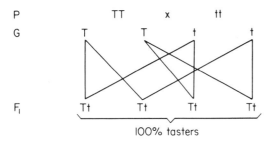

100% tasters

4. If a heterozygous taster marries a non-taster, the children should be 50% tasters and 50% non-tasters.

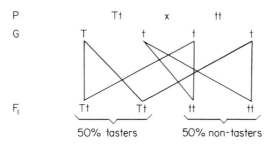

50% tasters   50% non-tasters

5. If two heterozygous tasters marry, the children should be 75% tasters and 25% non-tasters.

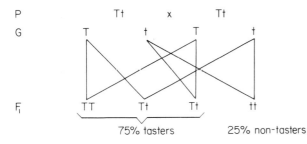

75% tasters   25% non-tasters

However, these percentages are only for very large numbers of offspring. Mendel had to breed many hundreds of pea plants to obtain his results. Hence 4. and 5. are unlikely to show these percentages in a normal family.

Another interesting human characteristic is the ability to curl up the sides of the tongue, usually referred to as 'tongue rolling'. The gene for tongue rolling is dominant and the genetical results that may be expected are the same as those for 'tasting'.

Sex Determination

Humans have 23 pairs of chromosomes in every cell of the body (except the spermatozoa or ova). In 22 of these pairs the two chromosomes are exactly alike. The 23rd pair are the sex chromosomes. These are alike in the female and are called X chromosomes. However, in the male there is one X chromosome and one Y chromosome. The Y chromosome has one end missing and is therefore slightly shorter than the X chromosome.

When ova are formed, each ovum will contain an X chromosome. When the spermatozoa are formed, half will contain an X chromosome and the other half will contain a Y chromosome. When mating occurs the ovum will have an equal chance of being fertilised by a spermatozoon containing an X or a Y chromosome. Consequently there are equal numbers of boys and girls born.

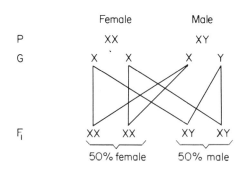

Sex-linked Genes

As with all chromosomes the sex chromosomes contain genes. The characteristics controlled by these genes will therefore be closely associated with the sex of the individual. Examples of such characteristics are haemophilia and red-green colour blindness.

Haemophilia is due to a lack of the gene causing the clotting of blood. It is a sex-linked disease which occurs only in the recessive state, that is when there is no dominant blood-clotting gene present. Females possess two genes, one on each X chromosome, but males only have one, as part of the Y chromosome is missing.

If we represent 'no haemophilia' by H and 'haemophilia' by h we can work out the possible genotypes that could occur.

The genotype of a female could be:

HH—no haemophilia.

Hh —no haemophilia, but she will have a 50% chance of passing the h gene on to her offspring.

hh —these females do not occur, presumably because their development is curtailed before birth.

The genotype of a male could be:

H —no haemophilia.

h —haemophilia. In these cases the child usually bleeds to death at an early age.

From this it can be seen that although females do not suffer from the disease, they may well be carriers, and can pass it on to their children. Fortunately it is rare.

Red-green colour blindness is the term given to the inability to distinguish red and green, both colours appearing as shades of grey. The gene for normal colour vision is dominant to that for red-green colour blindness and once again they occur on the sex chromosomes. These genes are transmitted in the same way as the haemophilia genes and produce the same results in that it occurs in men, and can be transmitted by females who are carriers. However red-green colour blindness differs from haemophilia in that the recessive state is not lethal, and so females can also be red-green colour blind. However, this occurs much more rarely than in men.

Chapter Twelve

FOOD

To be healthy a person requires in his food:

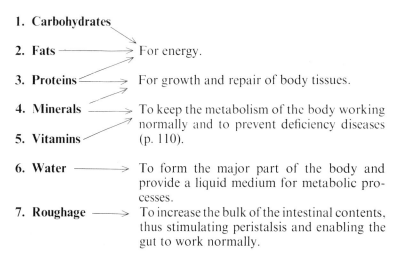

1. **Carbohydrates**

2. **Fats** ──────⟶ For energy.

3. **Proteins** ──────⟶ For growth and repair of body tissues.

4. **Minerals** ──────⟶ To keep the metabolism of the body working normally and to prevent deficiency diseases (p. 110).

5. **Vitamins**

6. **Water** ──────⟶ To form the major part of the body and provide a liquid medium for metabolic processes.

7. **Roughage** ──────⟶ To increase the bulk of the intestinal contents, thus stimulating peristalsis and enabling the gut to work normally.

CARBOHYDRATES

Carbohydrates contain the elements carbon, hydrogen and oxygen, with the hydrogen and oxygen in the same proportions as in water. This group includes sugars, starch and cellulose.

Sugars

(i) *Monosaccharides,* consisting of a single sugar molecule which has the general formula $C_6H_{12}O_6$, the arrangement of the atoms within the molecule determining the type of sugar.

e.g. **Glucose** which occurs in fruit juices
Fructose which occurs in honey and fruit juices.

(ii) *Disaccharides* $(C_{12}H_{22}O_{11})$, consisting of two mono-saccharides linked together chemically (minus a water molecule—H_2O).

e.g. **Sucrose** is a chemical combination of a glucose molecule and a fructose molecule. It occurs in sugar beet and sugar cane, fruits and carrots.
Maltose is a chemical combination of two glucose molecules. It is known as malt sugar as it is concerned in the germination of barley—a process known as malting—during the making of beer. In the body it is important as an intermediate stage in the breakdown of starch to glucose.

Starch

Starch is a **polysaccharide** $(C_6H_{10}O_5)_n$ being composed of large numbers of glucose molecules linked together chemically (minus water molecules).

(i) *Plant Starch* is produced from the glucose which is manufactured during photosynthesis. It is the main carbohydrate in food, occurring in flour, potatoes, rice and cereals.

(ii) *Animal Starch* is called **glycogen** and is built up in the liver and muscles from glucose. It is unimportant as a source of carbohydrate.

Cellulose

Cellulose is a polysaccharide and forms the cell walls of plants. It has no food value to man as it cannot be digested but it is very useful in the diet as roughage which aids peristalsis.

Function of Carbohydrates

Carbohydrates are a source of energy which is required for all the living processes. Much of this energy is given off as heat which enables the temperature to be maintained at about $37\,^{\circ}C$ ($98\cdot4\,^{\circ}F$). Excess carbohydrate can be stored in the body as glycogen, or converted into fat for storage.

FATS

Fats contain the elements carbon, hydrogen and oxygen, but in different proportions to carbohydrates. Fat is a compound of **fatty acids** and **glycerol,** usually 3 fatty acid units are combined with 1 glycerol unit to form a fat molecule. Fat can either be in a liquid state such as olive oil and corn oil, or in a solid state such as butter, margarine and lard. It also occurs in other foods such as cheese, meat, egg yolk and herrings.

Function of Fats

Fats are a source of energy, with a greater energy value than carbohydrate. Excess fat is stored in the body in a liquid state under the skin or around the organs.

PROTEINS

Protein contains the elements carbon, hydrogen, oxygen, nitrogen, and sometimes sulphur and phosphorus. A protein molecule is very large and complex and consists of several hundred **amino acid** units. There are about 20 different amino acids commonly found in food and they are present in varying amounts and arrangements in different proteins. Although the body can convert some of these from one sort to another, there are ten that must be included in the diet, and these are called **essential amino acids.** To provide these a variety of proteins are necessary in the diet, and can be obtained from meat, fish, eggs, milk, cheese, nuts, peas, beans.

Function of proteins

Proteins are essential for growth and repair of body tissues. Excess amino acids that are not required to build body proteins cannot be stored. They lose their nitrogen, which is excreted in the urine, and the remainder (consisting of carbon, hydrogen, oxygen) can be used for energy or converted into glycogen or fat for storage.

MINERALS

Minerals are the inorganic elements derived from food. These elements are required in relatively small amounts and some, called **trace elements,** are only needed in minute amounts, although all must be present in order to prevent deficiency diseases.

There are about 20 essential minerals and they tend to be widely distributed in food. Some examples are given in the table below. Note: care must be taken to ensure that iron and calcium in particular are present in adequate amounts. For this reason government regulations require that all flours must contain minimum quantities of these two minerals.

ELEMENT	NECESSARY FOR	MAIN SOURCES
Iron	Manufacture of haemoglobin.	Meat, eggs, vegetables, bread.
Calcium	(a) Bone and teeth formation. (b) Clotting of blood.	Milk, cheese, bread, green vegetables.
Phosphorus	Bone and teeth formation.	Most foods.
Iodine	Prevention of simple goitre (p. 78).	Usually in drinking water. Iodised table salt. Sea fish.
Fluorine	Prevention of tooth decay.	Some drinking water. May be added as fluoride to the water supply.
Sodium	(a) Sodium bicarbonate of pancreatic juice. (b) To help maintain correct composition of blood.	Most foods. Table salt (sodium chloride).
Chlorine	(a) Hydrochloric acid of stomach. (b) To help maintain correct composition of blood.	Most foods. Table salt (sodium chloride).

VITAMINS

Vitamins are complex organic compounds essential for life which the body is unable to manufacture for itself. They are only required in very small quantities but all are essential to allow the body to function properly and to prevent deficiency diseases. Vitamins can be divided into two groups according to whether they are soluble in fat or in water.

Fat Soluble Vitamins

	NECESSARY FOR	MAIN SOURCES
Vitamin A (can be stored in the liver for some months. An excessive intake can be harmful)	(a) Growth. (b) Formation of visual purple to prevent night-blindness. (c) Prevention of dry skin.	Butter, margarine, liver, green vegetables, carrots.
Vitamin D (can be stored in the body, an excess can have ill-effects in infants)	Prevention of rickets in children. (Lack of Vitamin D prevents calcium and phosphorus from being deposited in the matrix of bone tissue to give it strength.)	Margarine, fish oils, eggs, butter. Also formed in skin when exposed to sunlight.

	NECESSARY FOR	MAIN SOURCES
Vitamin E (Tocopherol)	Normal metabolism, although its function is not yet understood.	Many foods.
Vitamin K	Clotting of blood.	Manufactured by bacteria which live in the intestine and therefore rarely deficient. It also occurs in green vegetables.

Water-Soluble Vitamins

These cannot be stored in the body and therefore a regular supply in the diet is essential.

Note: The Vitamin B group is now known to consist of about 12 different substances.

	NECESSARY FOR	MAIN SOURCES
Vitamin B$_1$ (Thiamine)	Prevention of beri-beri (muscle weakness and nervous disorders).	Bread,* flour,* meat, potatoes, milk.
Nicotinic Acid (Niacin)	Prevention of pellagra.	Meat, bread,* flour,* vegetables, fruit, milk.
Vitamin B$_{12}$	(a) Red blood cell formation. (b) Prevention of degeneration of nerve cells.	Liver, meat, milk, eggs, fish.
Vitamin C (Ascorbic acid)	(a) Growth in children. (b) Prevention of scurvy (internal bleeding under the skin and swollen bleeding gums).	Fresh fruit and vegetables.

WATER

Water (H_2O) is essential for life and must be a daily part of the diet. It is the main substance in the body accounting for about 60–70% of the body weight, and it also provides a liquid medium in which the metabolic processes can take place. The water content of the body is maintained at a more or less constant level by the kidneys.

AVERAGE DAILY INTAKE		AVERAGE DAILY OUTPUT	
Drink	1300 cm^3	Urine	1500 cm^3
Food	850 cm^3	Breath	400 cm^3
Bi-product of tissue		Sweat	500 cm^3
respiration (p. 37)	350 cm^3	Faeces	100 cm^3
	2500 cm^3		2500 cm^3

2500 cm^3 = approx. $4\frac{1}{2}$ pints.

ROUGHAGE

This consists largely of cellulose and other indigestible plant material. As the human alimentary canal cannot digest roughage it has no food value, but it is useful in that it increases the bulk in the intestine and this stimulates peristalsis which enables the intestine to work normally. Main sources are vegetables and fruit.

CALORIES

The energy value of food is measured in terms of heat units called kilocalories (kcal), 1 kcal being the amount of heat required

---

* Thiamine (Vitamin B$_1$) and Nicotinic Acid (Niacin) are added to white flour by government regulations.

to raise the temperature of a kilogram of water by 1 °C. In everyday usage a kilocalorie is just called a calorie. The joule, the SI unit for measuring energy, is increasingly replacing the calorie. 1 kcal = 4·184 kJ (kilojoules) or approx. 4·2 kJ. 239 kcal = 1000 kJ = 1 MJ (megajoule).·

Different foods have different calorific values, that is, they provide different amounts of energy.

1 g carbohydrate produces		17 kJ (4 kcal)
1 g fat	,,	38 kJ (9 kcal)
1 g protein	,,	17 kJ (4 kcal)
1 g alcohol	,,	29 kJ (7 kcal)

It will be seen that fat has the highest calorific value and that alcohol is also an energy food. Much of the protein will be used for growth and repair and not for energy.

BALANCED DIET

This is one which contains all the food substances in the correct proportions to keep the person in a state of good health. The food requirements of the person will vary according to age, weight, sex, occupation, exercise taken, state of health etc.

Some examples giving average requirements are outlined below.

Infant—3·3 MJ (800 kcal)

The mother's milk contains all the food substances necessary for the first few weeks of life, except iron, but the baby usually is born with several months supply of iron stored in the liver. After a few weeks it needs to be supplemented with extra vitamins A, D and C. The same applies if the baby is being fed on cow's milk except that it will need more Vitamin C. Dried milk often has added vitamins and minerals and it is important to read the information on the label. After a few months, milk is gradually supplemented with other foods and by the time the child is a year old it will be eating a mixed diet similar to the rest of the family.

School Child—Age 8—8·4 MJ (2000 kcal)

Because the child is growing fast he will need extra protein, calcium and vitamins A, C and D. Also because he will be very active he will need a substantial amount of carbohydrate and fat for energy. Although a child of eight is not very large he requires almost as much food as an inactive adult, and he also must eat a variety of foods containing protein in order to obtain all the essential amino acids.

Adolescent Girl—9·6 MJ (2300 kcal) Boy 11·7 MJ (2800 kcal)

At this age boys are generally larger and more active than girls and therefore require more food. But the diet of both needs to be balanced in the same way as that for the eight-year-old. Adolescents have good appetites and care should be taken that too much carbo-hydrate and fat should not be eaten as in some cases this causes obesity (fatness) and skin disorders.

A school dinner should be a well-balanced meal providing on average 3·7 MJ, with 18·5 g animal protein, 32 g fat, and suitable amounts of vitamins and minerals.

Adult

The amount of kcal required will depend on the size of the individual, occupation, and amount of exercise.

Man

1. who has an active job or plays much sport. 15·1 MJ (3600 kcal)
2. who has a sedentary job, takes little exercise. 10·9 MJ (2600 kcal)

Woman

1. who leads a very active life or plays much sport. 10·5 MJ (2500 kcal)
2. most occupations. 9·2 MJ (2200 kcal)
3. who is pregnant. 10 MJ (2400 kcal)
4. who is breast-feeding. 11·3 MJ (2700 kcal)

Pregnancy requires an increased intake of protein, extra vitamins, particularly D and C, and extra calcium and iron.

Lactation (breast-feeding) requires a similar diet to pregnancy but with even more protein, and also extra Vitamin A and substances of the Vitamin B complex.

Old Person—8 MJ (1900 kcal)

Because less energy is used there should be less carbohydrate and fat in the diet. Care should be taken that this group has a balanced diet as there is a tendency for them to eat insufficient protein, fruit and vegetables.

Some of the terms which arise when talking about food and diet need definition.

Undernourishment occurs when insufficient food is taken into the body to keep it healthy. The extreme stage of this is starvation.

Malnutrition occurs when the quantity of food is adequate but the diet is unbalanced with too much of some food substances and not enough of others. Frequently there is too much carbohydrate and insufficient protein, minerals and vitamins.

Obesity is caused by an excessive calorie intake and the excess is converted into fat and stored, causing the body to become too fat. This can be the cause of much ill-health.

Hunger is a sensation which is felt when the stomach is empty and there is a desire for food.

Appetite is the desire for food which is not related to the body's need for it. Appetite can be due to eating habits or the sight or smell of tasty food.

FOOD TESTS

STARCH

(a) Stains blue-black with iodine solution.
(b) Is insoluble in water.
(c) Becomes sticky when heated with water.

SUGAR

(a) Tastes sweet.
(b) Is soluble in water.

Reducing Sugars (e.g. **glucose** and **fructose**) can reduce copper(II) sulphate to copper(I) oxide.

Non-reducing Sugars (e.g. **sucrose**) need to be boiled with dilute hydrochloric acid before they can do this.

Benedict's Test

This distinguishes between reducing and non-reducing sugars.

Make a solution of the substance to be tested. Put a little in a test-tube and add enough Benedict's solution (about 2 cm^3) to give a light blue colour. Boil carefully for one minute and if the colour changes to yellow, orange or red, the substance under test will be a reducing sugar. If the colour remains light blue, test to see if it is a non-reducing sugar. Put a little of the solution being tested in a test-tube, add a little dilute hydrochloric acid and boil for two minutes. Leave to cool. Neutralise by adding a little sodium bicarbonate until the fizzing stops, then add enough Benedict's solution to give a light blue colour and boil for one minute. If the colour now changes to yellow, orange or red the substance being tested will be a non-reducing sugar.

PROTEIN

(a) Coagulates (hardens) with cooking.
(b) Turns pink when heated with Millon's Reagent.

Millon's Test

In a test-tube make a solution or suspension of the substance under test. Add a few drops of Millon's reagent *(taking care, as this is poisonous)* and heat gently. Protein is present if the contents of the test-tube turn pink.

FAT

(a) Is greasy to the touch.
(b) Is insoluble in water and, in the liquid state, floats on top.
(c) Causes grease marks on paper. If the piece of paper is held up to the light the grease mark appears translucent.
(d) Dissolves in ether. If ether is added to a little olive oil and shaken the oil dissolves.
(e) In the liquid state stains red with Sudan III.

Sudan III Test

Put equal quantities of olive oil and water in a test-tube. Add a little Sudan III which is a red dye. Shake well and leave to settle. The fat takes up the red stain and floats on the water which has been left colourless.

PRESERVATION OF FOOD

REASONS FOR STORAGE

1. As food production is mainly seasonal, it is necessary to store it after the harvest for use during the rest of the year.

2. In order to transport it from the place where it is produced to the place where it is to be consumed. Britain does not produce

nearly enough food to feed the nation and therefore imports food from all over the world.

3. To have a more varied choice of food which allows a balanced diet, and is also good for the appetite.

4. Convenience. These days when housewives lead more active lives outside the home it is convenient to buy food already prepared for cooking or eating. This is done in factories and is then preserved by canning, quick-freezing or dehydrating, and stored until required.

As our food comes from plant or animal matter it is natural for it to go bad (decompose) because:

1. Micro-organisms, particularly bacteria and fungi, will live on it.

2. Enzymes within the plant or animal cells continue to be active.

Therefore, if food is to be stored it needs to be preserved to stop or delay the activity of micro-organisms and enzymes.

Food preservation includes any method of treating food so that it retains all, or most, of its nutritional value and can be stored for use at a later date. Today, the important methods of food preservation are canning, quick-freezing and dehydration.

CANNING

Canning was the first method of food preservation to become important and practically every kind of food can be preserved in this way—meat, fish, vegetables, fruit, milk, puddings, drinks. The food is put into a tin can, the air is removed, a vacuum created, and the can is sealed. It is then sterilised at a high temperature to destroy the micro-organisms and enzymes. This also cooks the food.

Advantages of Canning

1. Food is sterilised.
2. The food value is not much affected.
3. Cans are easy to transport and store.
4. The food remains edible for several years, meat having a longer storage life than fruit because acids damage the can in time.

Disadvantages of Canning

1. Costly to transport.
2. Bulky to store.
3. Food is always in the cooked state.

QUICK-FREEZING

Quick-freezing was initiated by Mr Birdseye in America in the 1920's. The food is frozen as rapidly as possible below a temperature of $-18\,°C\,(0\,°F)$ which reduces the activity of micro-organisms and enzymes. The quicker the food is frozen the better because:

1. The water in the food will not have time to form large ice crystals, and these cause greater damage than small ones.

2. Less damage is done to protein which, therefore, retains more of its nutritional value.

3. It reduces the time in which micro-organisms and enzymes can act to decompose it.

4. It is more economical as the costly machinery can deal with a larger volume of food.

Frozen food must be packed in polythene or other moisture and vapour proof material. This is necessary to prevent water escaping from the food (causing it to dry up) and air reaching it (encouraging decomposition).

Once frozen, the food must be stored at a temperature below $-18\,^{\circ}C\ (0\,^{\circ}F)$. If it thaws this will allow the bacteria and enzymes to become active again. It is then dangerous to re-freeze the food in case harmful bacteria have been growing and multiplying in it. Although they will be inactivated by the cold, the large numbers now present will quickly become active again as soon as the food is thawed before being eaten. They will then either destroy the food or, much worse, cause food poisoning if eaten.

Advantages of Freezing

1. Food can be preserved uncooked.

2. As food needs to be frozen quickly it will retain its freshness.

3. It is only worthwhile freezing top quality food. Therefore, frozen food will be good food.

4. Meals can be cooked in advance and frozen until required which makes catering easier.

Disadvantages of Freezing

1. Frozen food can only be transported in refrigerated trucks and then must be stored in the shop or home in deep-freeze cabinets.

2. It does not have a very long storage life, ranging from a few weeks for shell-fish to about a year for fruit, vegetables, beef and poultry. At the end of its storage life the food may remain edible but the quality deteriorates.

3. Once the food has thawed it must be eaten within a fairly short time.

DEHYDRATION

Dehydration involves removing most of the water from the food and this prohibits the action of micro-organisms and enzymes. Although drying of some foods, e.g. grain, has been done for thousands of years it is only recently that techniques have been developed for dehydrating a much wider range of food, e.g. vegetables, soups, milk, complete meals. It is now possible to dehydrate them in such a way that when they are rehydrated they are both edible and palatable.

Hot Air Drying

This is the usual method, water from the food evaporating into the hot air.

Vegetables can be dried by the hot air passing up through perforated moving trays which prevents the pieces sticking together and also accelerates the drying time.

Milk and coffee can be sprayed into hot air and dried as powder.

Instant potato and breakfast cereal can be spread over hot rollers and dried as small flakes.

Accelerated Freeze-Drying (AFD)

In this process the food is first frozen solid and then the ice is caused to evaporate. This removes more moisture and as it is a gentler process it alters the structure of the food less. But it is a more costly process and the food becomes fragile. It needs careful packaging to prevent it crumbling, and it also needs to be vacuum packed or packed with nitrogen or carbon dioxide to prevent oxidation. It is used mainly for expensive foods, e.g. chicken and prawns, which do not dry satisfactorily by the hot air method.

Advantages of Dehydration

1. The food is very light in weight and, therefore, cheaper to transport.
2. If packed correctly it is easy to store.
3. Has a long storage life.

Disadvantages of Dehydration

1. Before it can be eaten it has to be rehydrated, that is, it has to absorb water. During dehydration the food must be processed so that it can be rehydrated successfully.
2. If the food has been dried by AFD it will be expensive, and it also requires costly packaging.

OTHER METHODS OF FOOD PRESERVATION

Pasteurisation of milk. The milk is heated at 71–73 °C for about 15 seconds then cooled rapidly. This kills most of the bacteria, does not alter the taste and prolongs the storage life of milk by several days.

Sterilisation of milk. The milk is heated to 140 °C for 1–2 seconds and sealed in air-tight bottles. This kills all the bacteria and allows the milk to be kept for several months, but it alters the taste.

Bottling of fruit. The fruit is sterilised in the bottles to kill bacteria and enzymes and then sealed to make it air-tight.

Curing of bacon and fish. This can be done by smoke or salt and in both cases the food is dehydrated to such an extent that micro-organisms cannot grow.

Pickling of vegetables and eggs. They are preserved in vinegar which makes them too acid for the activity of micro-organisms or enzymes.

Refrigeration The food is kept at temperatures just above freezing point which reduces the activity of micro-organisms and enzymes. This method is suitable for short-term storage only.

FOOD CONTAMINATION

Food which is contaminated can cause food poisoning. Food is said to be contaminated if it contains harmful bacteria or their products, called toxins, or harmful chemicals, all of which can give rise to illness.

Food poisoning can be divided into several types:
1. Caused by poisons which occur naturally, for example those

which occur in the deadly nightshade plant and in certain toadstools.

2. Due to sensitivity of a particular individual to some foods. This produces an allergy reaction, for example the 'nettle rash' reaction after eating strawberries or shellfish.

3. Chemical poisoning caused by chemical substances accidentally being either included in the food or derived from the container. For example, acid fruit drinks can dissolve zinc, lead and antimony from utensils of inferior quality. Of more importance these days is the accidental contamination of food by an insecticide or pesticide.

4. Bacterial poisoning caused either by bacteria or their toxins (poisons). This is the most usual type of food poisoning, the characteristic symptoms being vomiting, diarrhoea and abdominal pain. The bacteria come into contact with the food, grow and multiply under the right conditions, and are then eaten. The right conditions for the growth of bacteria include warmth and moisture, and sufficient time to multiply.

Bacteria which cause food poisoning:

(a) **Salmonellae**   There are many different types of Salmonellae, and they usually live in the bowel. They can occur in meat, meat products, soups, milk and eggs. Thorough cooking will kill them and thus prevent illness.

(b) **Staphylococci**   These are very common and are found almost everywhere, including in the nose and throat and in boils and the pus from an infected wound. They are able to grow, multiply and produce much toxin in foods such as meat, custards and cream cakes, if these foods are kept at a warm temperature. If eaten, the toxin is the cause of the illness. Although the bacteria are killed if heated, the toxin is heat-resistant and is not destroyed by cooking.

(c) **Clostridia**   Two species of *Clostridium* are important in food poisoning. Both thrive in the absence of oxygen and produce heat-resistant spores, which can survive normal cooking processes.

Clostridium welchii thrives in such foods as warm gravy and stews and is a common form of food poisoning. The slow cooling or re-warming of meat dishes can lead to rapid production and multiplication of bacteria from the spores. The illness is generally of short duration and, although unpleasant, it is not usually serious.

Clostridium botulinum produces a toxin which gives rise to a serious and often fatal form of food poisoning (botulism). If canned meat or vegetables are not properly preserved, the spores of this bacterium can survive and germinate in the anaerobic conditions within the can. Botulism from commercially canned foods is rare.

FOOD HYGIENE

Contamination of food can be prevented by:

1. Personal cleanliness of those who handle food in the factory, shop and home. Hands should be washed, particularly after visiting the lavatory. Touching the nose and mouth, licking the fingers and coughing and sneezing over food should all be avoided. Boils and septic cuts should be covered. Clothing should be clean. Any person with a stomach upset or diarrhoea should avoid handling food.

2. Cleanliness of factories and shops handling food.
3. Cleanliness of kitchen utensils, crockery and cutlery.
4. Correct methods of preservation.
5. Correct methods of storage.
6. Keeping flies, mice and other pests away from food.
7. Sufficient cooking of food, when necessary.
8. Re-heating cooked foods thoroughly.
9. Using clean water; either purified tap water or boiled water.

In addition there are regulations governing the misuse of chemical additives to food.

FOOD INSPECTION

The Food and Drugs Act 1955 and various other Acts of Parliament have been passed to safeguard our food supplies. The Environmental Health Officers help to enforce these regulations, and food producers, manufacturers and retailers can be prosecuted for failure to maintain the correct standards. Should an outbreak of food poisoning occur, the local health authority must be notified and an investigation is carried out to try to determine the cause.

Food Hygiene Regulations are in force concerning the cleanliness of food premises, equipment, storage, transport and personnel of the food industry. If food handlers are infected with germs which can cause food poisoning or other food-borne diseases, the local health authority must be notified.

At slaughter houses an examination of meat is made by a veterinary surgeon or Environmental Health Officer. If it is diseased or otherwise unfit for human consumption it has to be specially treated and can only be sold as pet food. E.H.O.'s also carry out routine inspections of dairies and ice cream factories and send samples of milk and ice cream to the Public Health Laboratory. Inspections are also made, and samples taken from other food premises, such as food factories, greengrocers, bakers, fishmongers, public houses, cafés etc. The samples of food and drink are sent to the Public Analyst to check that they are fit for human consumption and are properly labelled. Any food or drink liable to cause ill-health must be destroyed. If any labels are incorrect, these must be put right.

Chapter Thirteen

DISEASE

Disease is present in the body when part of it does not function properly. There are a variety of causes.

INHERITED DISEASES

Inherited diseases are caused by genes inherited from parents. They can only be transmitted from parent to offspring.

Examples

Haemophilia (p. 97).

Phenylketonuria (PKU)—causes mental deficiency. Newborn babies are given a routine blood test (it used to be a urine test) to detect the condition because, if children with PKU are fed on a special diet, the brain can develop normally.

CONGENITAL DISEASES

Congenital diseases are caused by the imperfect development of the baby while it is in its mother's womb.

Examples

Thalidomide babies—a drug taken as medicine by the mothers prevented normal growth of the limbs.

German measles—if the mother suffers from this disease at a certain stage of pregnancy, the baby may have defects of the eyes, ears or heart.

Congenital syphilis—if a pregnant mother has syphilis, her child may acquire it.

DEFICIENCY DISEASES

Deficiency diseases are caused by the lack of certain substances in the diet.

Examples

Lack of Vitamin B_{12} causes malformation of the red blood cells, resulting in a disease known as pernicious anaemia.

Lack of Vitamin C causes scurvy.

Lack of Vitamin D causes rickets in children.

Lack of iodine causes simple goitre.

Lack of iron reduces the formation of haemoglobin and results in iron-deficiency anaemia.

INDUSTRIAL DISEASES

Industrial diseases are those caused by the working conditions in particular industries.

Examples

Coal dust can cause pneumoconiosis in coal miners.

Silica dust can cause silicosis in quarry workers.

Asbestos dust can cause asbestosis in workers manufacturing asbestos goods. These examples are all lung diseases.

Prolonged use of pneumatic drills can cause arthritis and nerve damage to hands and forearms.

Mercury used in the manufacture of felt hats can cause 'hatter's shakes', a form of chronic mercury poisoning.

Lead can cause poisoning in painters.

ENVIRONMENTAL DISEASES

Environmental diseases are those caused by conditions of the environment.

Examples

Tuberculosis may be due to adverse environmental conditions such as overcrowding which favour the spread of the disease.

Rheumatic diseases are often associated with damp living conditions.

Goitre is associated with areas deficient in iodine.

INFECTIOUS DISEASES

Infectious diseases are those caused by pathogens which invade the body and grow and multiply in the tissues.

Pathogens

Pathogens are parasites which cause disease, they are often referred to as **germs.** (**Parasite**—an organism that lives on or in another living organism, the host, from which it obtains food.) Most pathogens are micro-organisms, commonly viruses, bacteria, protozoa and fungi.

Viruses

Viruses are so small that they cannot be seen with an ordinary microscope but need the use of an **electron microscope.** They show only some of the properties of life and can be crystallised into what appears to be a non-living form. However, they can only live and reproduce within living cells, and cause a number of diseases in plants and animals. After infecting a host cell, the virus uses the cell to produce new virus particles. The host cell is usually killed in the process and the virus particles are released to infect other cells. Unlike most bacteria they cannot be grown on culture media but only in living cells such as kidney tissue or incubated fertilised hen's eggs. This type of culture is referred to as **tissue culture.**

Infection by viruses results in diseases such as the common cold, influenza, poliomyelitis, etc.

Bacteria

Bacteria are minute organisms which are found everywhere, but it is impossible to see them without the aid of a microscope.

111

Conditions necessary for the growth of bacteria and their importance to man are discussed in Chapter 1 (p. 6).

Each bacterium is a single cell but it lacks a definite nucleus and chlorophyll. Substances necessary for life are absorbed through the cell wall and unwanted substances are excreted by the same route. After growing to a certain size it then reproduces by dividing into two. Under favourable conditions this can happen about every 20 minutes and so large numbers can be built up very quickly. It is possible to grow bacteria artificially on culture media in order to study them (see page 117). Bacteria will be seen to develop on the culture medium such that in the course of 24 hours 1 bacterium will multiply into many millions resulting in a 'colony' visible to the naked eye. Some bacteria are capable of independent movement by means of small hair-like processes called **flagella**.

Bacteria pathogenic to man thrive best at about blood heat (37 °C or 98·4 °F). The symptoms of disease are either caused by the bacteria or by their waste products, which may be toxic (poisonous) to the host.

Bacteria are classified according to shape.

Cocci are spherical.

Some occur singly.

Some occur in pairs—*diplococci* (cause gonorrhoea, meningitis).

Some form chains—*streptococci* (cause tonsillitis, scarlet fever).

Some form clusters—*staphylococci* (cause boils, food poisoning).

Bacilli are rod-shaped.
Cause typhoid, tetanus, tuberculosis, diphtheria.

Vibrios are curved (comma-shaped).
Cause cholera.

Spirochaetes are twisted like a cork-screw.
Cause Weil's disease, syphilis.

Protozoa

Protozoa are minute unicellular animals, each having a nucleus.
Most are harmless to man, but a few are pathogenic.
Amoebic dysentery is caused by **Entamoeba histolytica.**
Malaria is caused by various species of **Plasmodium.**

Fungi

Fungi are simple plants which lack chlorophyll. The group includes yeasts, moulds and mushrooms. A few microscopic fungi can cause skin diseases including ringworm, athlete's foot and thrush. Fungi and their importance to man are discussed in Chapter 1 (p. 8).

112

TRANSMISSION OF DISEASE

Infectious disease can be spread by:

1. Contact—by touching an infected person or an article that has been in contact with an infected person. Diseases spread in this way are called **contagious** diseases, and this group includes many skin diseases, e.g. scabies, ringworm and smallpox.

2. Droplet Infection Diseases of the respiratory tract are often spread in this way, e.g. the common cold and influenza. When an infected person speaks, coughs or sneezes minute droplets containing germs are sent into the air which may then be inhaled by another person.

Fig. 13.1 Life history of the housefly and its role in spreading disease.

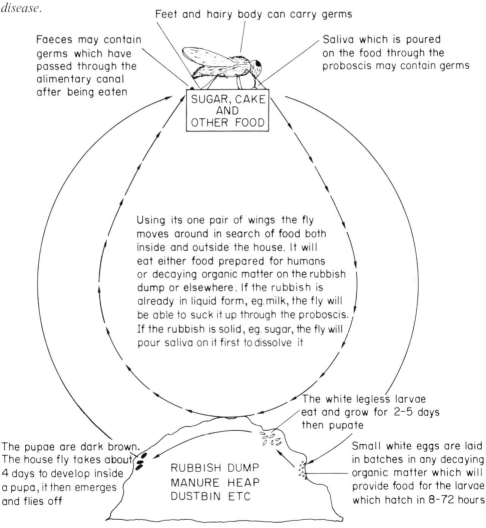

Feet and hairy body can carry germs

Faeces may contain germs which have passed through the alimentary canal after being eaten

Saliva which is poured on the food through the proboscis may contain germs

SUGAR, CAKE AND OTHER FOOD

Using its one pair of wings the fly moves around in search of food both inside and outside the house. It will eat either food prepared for humans or decaying organic matter on the rubbish dump or elsewhere. If the rubbish is already in liquid form, eg. milk, the fly will be able to suck it up through the proboscis. If the rubbish is solid, eg. sugar, the fly will pour saliva on it first to dissolve it

The white legless larvae eat and grow for 2–5 days then pupate

The pupae are dark brown. The house fly takes about 4 days to develop inside a pupa, it then emerges and flies off

RUBBISH DUMP MANURE HEAP DUSTBIN ETC

Small white eggs are laid in batches in any decaying organic matter which will provide food for the larvae which hatch in 8–72 hours

Diseases carried by the house fly include infectious diarrhoea, typhoid, dysentery. Prevention (1) kill flies in the house (2) keep food and cooking utensils covered (3) treat dustbins etc. with chemicals to kill eggs, larvae and pupae (4) Keep rubbish dumps covered so that the heat generated by the decaying organic matter will kill the eggs, larvae and pupae

3. Contaminated Food If food becomes contaminated during its preparation by germs from an infected person, or flies, or dust and dirt, it may give rise to disease of the digestive tract. Examples of **food-borne** diseases are typhoid and salmonella food poisoning.

4. Contaminated Water Intestinal diseases can also be spread by drinking contaminated water. Some bacteria can live for days or even weeks in water. The chief source of contamination is sewage and this is why Sewage Disposal and the Water Supply must be kept separate. Untreated water should be boiled before drinking to prevent such water–borne diseases as typhoid, cholera and amoebic dysentery.

5. Human Carriers These are people who are infected by a pathogen but not affected by it (they have an immunity to it). They may not be aware that they are harbouring the disease and mix freely with other people, consequently they may pass on the disease, e.g. typhoid, diphtheria.

6. Insect Vectors These are insects which spread disease.
   (i) By carrying the germs externally on their feet and bodies, e.g. houseflies carry germs from rubbish dumps to food (Fig. 13.1).
   (ii) By carrying the germs internally and transmitting them to the human when they pierce the skin to suck blood. In this way malaria is spread by the mosquito, sleeping sickness is spread by the Tsetse fly and typhus by the louse and the rat flea.

7. Animal Vectors These are other animals which spread disease, e.g. the rat, which can spread Weil's disease, plague and typhus.

Rabies is spread by foxes, dogs, cats and other mammals (p. 129).

DEFENCES OF THE BODY AGAINST INFECTIOUS DISEASE

The body is continually under attack by germs from the environment but disease rarely results because:

1. The skin forms a barrier to the entry of germs. This also applies to blood clots over wounds.

2. Gastric juice contains hydrochloric acid, which kills germs before they enter the intestine.

3. Tears, mucus and saliva are all mildly antiseptic, as they contain enzymes called **lysozymes** which are capable of killing germs.

4. Lymph nodes filter the blood, and germs are destroyed there by the white blood cells.

5. White blood cells collect at the site of infection to destroy the germs. Repeated infection by a particular germ causes white blood cells to become specialised in recognising and dealing with this type of germ and they are referred to as **sensitised cells.**

6. The presence of **antibodies** in the blood destroys germs or makes them and their toxins harmless. Antibodies are protein substances produced by the body in response to **antigens,** an antigen being any foreign matter, e.g. germs, vaccines, that causes an immune response in the body. The production of antibodies in quantity takes about 10–14 days and gives the person immunity against disease, but they are very specific, one type of antibody affecting only one type of germ.

7. Pain makes the body aware of infection and encourages treatment.

IMMUNITY

Immunity is the ability of the body to resist infection. There are two main categories:

1. Natural Resistance

The body has a natural resistance to many germs. Either they are unable to penetrate into the tissues of the body or, if they do, the environment within the body is uncongenial to their survival, e.g. man's immunity to swine fever, distemper and myxomatosis.

2. Acquired Immunity

The body acquires freely circulating antibodies or sensitised cells which gives immunity and follows either contact with the disease, or the vaccination of antigens, or the injection of antibodies.

    (i) *Active Immunity* is when the body actively produces antibodies against the antigen. This gives long-term immunity.

        (a) *Natural Active Immunity* occurs following an attack of the disease, e.g. whooping-cough.

        (b) *Artificial Active Immunity* occurs following vaccination, e.g. smallpox vaccination.

    (ii) *Passive Immunity* is when the body receives antibodies which have been produced in another person or animal. This gives short-term immunity.

        (a) *Natural Passive Immunity* occurs when a child in the womb acquires antibodies produced by the mother, e.g. measles. This is also called **congenital immunity.**

        (b) *Artificial Passive Immunity* occurs following the injection of antibodies. Although this gives only short-term immunity it is useful if the person has been exposed to, or is suffering from, the disease, e.g. tetanus antitoxin. (**Antitoxin**—a type of antibody that can neutralise toxin.)

ORGAN TRANSPLANTS AND TISSUE REJECTION

Immunity is also the cause of tissue rejection which occurs when organs (e.g. heart, kidney, skin) are transplanted from one person to another. Organ transplants can easily take place between identical twins as they both have the same genetic make-up, and they sometimes succeed when a near relation donates the organ, e.g. when a father donates a kidney to his son. But the problems involved in transplanting organs between strangers are very difficult to solve. The patient's body regards the transplanted organ as foreign matter so the white blood cells invade it and eventually destroy the tissues. Drugs are used to combat this tissue rejection by immobilising the white blood cells. Unfortunately, this leaves the patient with much reduced resistance to infection and, despite all precautions, he may contract a harmful or even fatal disease.

VACCINATION

Vaccination, also called **immunisation,** is the process of administering vaccine to the body so that it can produce immunity against a specific disease.

The vaccine may be:

1. Dead pathogens, the germs being killed by heat or chemicals

before being injected, e.g. vaccines against influenza, typhoid, cholera, and the Salk vaccine against polio.

2. Toxoid, which is a harmless form of the toxin which the disease produces, e.g. vaccines against diphtheria and tetanus.

3. Live, attenuated vaccine. This consists of living bacteria or viruses which grow and multiply in the body without causing illness. They have been selected because they are closely related to the disease-causing organism and cause the body to produce the correct antibodies against the disease, e.g. vaccination with cowpox virus produces immunity against smallpox, BCG vaccine produces immunity against tuberculosis, the Sabin vaccine against polio, and measles vaccine.

INFECTION

Infection occurs when the disease gains control in the body.

Pattern of Infectious Disease

1. Germs enter the body.
2. Incubation—the germs grow and multiply.
3. Symptoms appear. Generally fever, headache, sickness, and others according to the disease.
4. Crisis—the severest stage of the disease ending in either death or recovery.
5. Convalescence—the symptoms disappear and the patient regains strength.

Treatment of Infectious Disease

This will vary according to the disease but the following treatments may be recommended.

1. Alleviate the symptoms and allow the body to produce its own antibodies to destroy the infection, e.g. common cold.
2. Inject antibodies or antitoxins to render the germs or their toxins harmless (e.g. diphtheria).
3. Use medicines to prevent the germs growing and multiplying, e.g.

   (i) **Antibiotics** These are substances produced by living organisms, usually fungi, which inhibit the growth of bacteria and other micro-organisms (e.g. *Penicillin*). Different antibiotics affect different micro-organisms. Unfortunately, new strains of disease are constantly appearing which are resistant to the antibiotic that previously cured the disease. So new antibiotics are in constant demand. They can be taken as tablets or in a liquid form, or can be injected, or applied as an ointment. Some patients become allergic to certain antibiotics, the commonest being penicillin, and this prohibits their future use in these patients.

   (ii) **Sulphonamides** These are a group of drugs containing sulphur which work in a similar way to antibiotics. They were discovered before antibiotics.

Epidemic

This occurs when a disease spreads rapidly and involves large numbers of people. It will be an infectious disease that can be transmitted easily, e.g. influenza. The epidemic will spread because conditions are favourable to the pathogen and its transmission. The epidemic will be over when conditions become unfavourable.

An epidemic becomes **pandemic** if it spreads to many countries throughout the world.

A disease is said to be **endemic** if it is always present in a particular part of the world, e.g. malaria in the tropics.

A disease is said to be **sporadic** if it occurs in different places at different times, with no known connection between the outbreaks.

Conditions Favourable to Epidemics
1. Little or no immunity of the population.
2. Virulent strain of the disease, e.g. Hong Kong 'flu' virus.
3. Means of transmission easy, e.g. for droplet infection large numbers of people confined together; for transmission by insect vector the presence of large numbers of the right insect.
4. Population with a lowered state of resistance to infection, e.g. in the winter, or due to malnutrition.

Conditions Unfavourable to Epidemics
1. Increased immunity of population, e.g. by vaccination.
2. Means of transmission reduced or stopped, e.g. open windows in summer reduces the chance of droplet infection; destruction of the insect vector stops the spread of malaria.
3. Population in a good state of general health and therefore greater resistance to disease, e.g. in summer, good nutrition.
4. Killing of germs by disinfectants, antiseptics, heat treatment, sunlight, etc.

Sterilisation Sterile or **aseptic** means free from microbes (micro-organisms). Complete sterilisation can be achieved by compressed steam in an **autoclave.** Most microbes are killed by **sunlight** (ultra-violet rays), also by **boiling water,** but some bacterial spores can withstand prolonged boiling. **Disinfectants** and **antiseptics** are chemicals which destroy microbes; generally, antiseptics are used on people and disinfectants are used on things (drains, apparatus etc).

EXPERIMENT TO SHOW THE PRESENCE OF MICRO-ORGANISMS IN AIR, WATER, MILK, AND ON THE HANDS

Method
Prepare and sterilise 5 petri dishes containing nutrient agar.

Fig. 13.2 Petri dish.

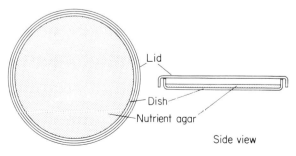

From above

117

Petri dishes are made of glass or plastic and each consists of two parts, one part is placed over the other to form a loosely-fitting lid (Fig. 13.2).

Nutrient agar is the culture medium and contains both food and water and provides a jelly-like base for the micro-organisms to grow on.

It is not too difficult to prepare and sterilise the dishes and agar for this experiment but it may be more convenient to purchase the 5 dishes already filled with nutrient agar and sterilised by ultra-violet light. These are readily obtainable from laboratory suppliers.

Dish 1. Leave lid off for five minutes.
Dish 2. Lift up lid to allow a drop of unboiled water to be placed quickly on the agar.
Dish 3. As for dish 2, but add a drop of milk instead of water.
Dish 4. Lift off lid, place fingers on the agar, remove quickly and replace lid.
Dish 5. Leave untouched in order to have a control (necessary in all biological experiments).

Leave the dishes in a warm dark place and examine daily.

Result

Dish 1 has colonies of bacteria and fungi scattered over the agar. Dishes 2, 3, and 4 have colonies of bacteria and fungi growing where the water, milk and finger prints respectively, touched the agar.

Dish 5. The control should remain free of micro-organisms. If it does show any growth this means that the dishes and agar were not properly sterilised and the experiment must be considered void.

Although both the milk and water contain micro-organisms these will be harmless if the milk has been pasteurised and the water purified in a Water Treatment Plant.

DETAILS OF SOME INFECTIOUS DISEASES

COMMON COLD

Cause	A large group of viruses.
Transmission	Droplet infection.
Incubation	About 2 days.
Symptoms	Fever, headache, over-secretion of the mucous membrane of the nose.
Treatment	Alleviate symptoms.
Prevention	None, except avoiding people with the infection.

INFLUENZA

Cause	A group of viruses.
Transmission	Droplet infection.
Incubation	About 2 days.
Symptoms	Fever, aching body, weakness, coughing. Secondary infections, such as bronchitis, sometimes follow.
Treatment	Warmth and rest.

Prevention   So far no vaccine has been discovered which guarantees immunity against influenza because of the variety of different strains of the disease. But vaccination in the autumn may give immunity to the more common strains over the winter months when influenza can be caught so easily in crowded atmospheres.

The virus is cultured in incubated fertilised hen's eggs. Amniotic fluid is then extracted which is injected into the human body causing it to produce antibodies against that particular strain of the virus. The vaccines now in use are a mixture of these fluids so that the body will produce antibodies against several strains of the disease.

MEASLES

Cause   Virus.
Transmission   Droplet infection or contact.
Incubation   About 10–14 days.
Symptoms   Fever, head cold, cough, rash. It is serious if it leads to complications in ears, eyes and lungs, etc.
Treatment   Usually none.
Prevention   (a) Vaccination of babies over one year old with live attenuated virus. Only one injection is needed. This is an example of artificial active immunity.
(b) Congenital immunity. A mother who has had measles passes immunity on to her child which lasts for about one year.
(c) Natural active immunity will be acquired by having the disease, a second infection being rare.

POLIOMYELITIS

Cause   Virus.
Transmission   Droplet infection or contact.
Incubation   Usually 7–12 days.
Symptoms   Fever and infection of nervous system which may cause paralysis of one or many muscles, and may be temporary, or permanent.
Treatment   Rest, and the use of an iron lung if the respiratory muscles are involved. Later, if necessary, massage and therapy.
Prevention   Vaccination of babies. Three doses of vaccine are given at monthly intervals with a booster dose at five years. The vaccine can be injected or given orally on a sugar lump.

Salk vaccine contains dead polio virus.
Sabin vaccine contains live attenuated virus.

SMALLPOX

Cause   Virus.
Transmission   Contact with the pus from the 'pocks', either by touching the person or contaminated articles.

Incubation	7–16 days.
Symptoms	High fever, rash, the 'pocks' (spots) become filled with pus, scabs form and leave scars when they fall off (pock-marks).
Treatment	Isolation in special hospitals.
Prevention	Vaccination with live attenuated virus. This is done by puncturing the surface of the skin and applying the vaccine.

In Britain, infants were commonly vaccinated but this is no longer generally recommended as the disease has now been wiped out in this country. However, it is advisable to be vaccinated if travelling to a part of the world where the disease is endemic, with revaccination every 3–5 years.

INFECTIOUS DIARRHOEA

Also called Epidemic Diarrhoea in Children, Summer Diarrhoea, Gastroenteritis, etc.

Cause	Various micro-organisms which are often difficult or impossible to identify.
Transmission	Food, water or milk contaminated by the faeces of an infected person or, by dust and dirt.
Incubation	Usually 1–4 days.
Symptoms	Diarrhoea, sometimes with sickness, pains, fever, malaise, etc.

Diarrhoea is the term given to abnormally frequent emptying of the bowels, causing the faeces to be expelled before food and water have had time to be absorbed. It can affect people of all ages and is particularly dangerous in the old and the very young because of the loss of water from the body, a condition known as dehydration.

Treatment	Replacement of water loss and sometimes antibiotics.
Prevention	General cleanliness, especially in the preparing and eating of food.

CHOLERA

Cause	Bacterium.
Transmission	Water contaminated by the excreta of infected persons. Also by contaminated food.
Incubation	A few hours to 5 days.
Symptoms	Diarrhoea and sickness, followed by cramp, thirst, exhaustion.
Treatment	Very important to replace the water lost from the body.
Prevention	(a) Purification of the water supply. (b) Vaccination with dead cholera bacteria.

DIPHTHERIA

Cause	Bacterium.
Transmission	Contact with an infected person.
Incubation	Usually 2–5 days.

Symptoms	Fever, sore throat and neck. Toxins from the bacteria may affect other organs of the body.

Symptoms Fever, sore throat and neck. Toxins from the bacteria may affect other organs of the body.

Treatment Injections of antitoxin to act against the toxins.

The antitoxin is obtained from a horse which has been injected with diphtheria toxin. Although it will not suffer, it will produce antitoxins in its blood. When the blood serum containing the antitoxins is removed, it can be used to reduce the severity of the disease in humans.

Prevention Immunisation of babies with toxoid. Three injections are given at monthly intervals with booster doses at five and eleven years. The vaccine is included in the combined vaccination against whooping-cough, diphtheria and tetanus.

Immunisation has almost wiped out the disease in this country.

TETANUS (LOCK-JAW)

Cause Bacterium.

Transmission Germs in soil, dust or dirt enter the body through wounds. Tetanus occurs particularly in farming country where supplies of manure are available.

Incubation 4–21 days.

Symptoms Stiffness and pain of jaw muscles, and sometimes other muscles.

Treatment Injection of antitoxin.

Prevention Injection of toxoid causes the body to produce antibodies. For babies, it is included in the combined vaccination against whooping-cough, diphtheria and tetanus. Three injections are given at monthly intervals. A booster dose for tetanus is needed at 5 and 11 years. A booster dose is also given if the person is injured. If, however, an injured person has not previously been vaccinated against tetanus, then an immediate injection of antitoxin is given.

TUBERCULOSIS (TB)

Cause Bacterium.

Transmission Droplet infection or contact.

Incubation 4–6 weeks.

Symptoms May attack various parts of the body but most commonly the lungs where it causes tissue destruction. Sometimes the infection may be so mild as to go unnoticed.

Treatment Specific antibiotics.

Prevention BCG vaccination (Bacillus Calmette-Guerin). The vaccine is a freeze-dried preparation containing live attenuated bacteria and is given to people not possessing sensitised cells against TB bacteria. Therefore, it is necessary to give the Tuberculin test first to see if the person has these sensitised cells due to either
(a) a past infection, or
(b) a current infection.

Tuberculin test A small drop of tuberculin (material from TB bacteria) is injected into the skin. If a red spot develops it shows that the person has the necessary sensitised cells already present in the blood. The test is said to be positive.

If the test is negative, a BCG injection is given and the necessary sensitised cells will be produced which will give immunity for 8–10 years. The usual age for the injection is at 13, as adolescents and young adults are more liable to the disease. Babies are only vaccinated if they have come into close contact with an infected person.

TYPHOID FEVER

Cause	Bacterium.
Transmission	Food and water contaminated by the excreta of infected persons. Sometimes flies act as vectors by carrying the germs from excreta to food.
Incubation	About 14 days.
Symptoms	Fever with constipation or diarrhoea.
Treatment	Antibiotics.
Prevention	(a) Vaccination with two injections of TAB vaccine given with an interval of not less than 7–10 days. TAB vaccine contains dead typhoid, paratyphoid A and paratyphoid B bacteria, and also gives immunity against paratyphoid fever (milder than typhoid).
	Vaccinations should be given before going to parts of the world where the disease is endemic, with a booster dose every 1–3 years.
	(b) Purification of the water supply.
	(c) Hygienic methods of sewage disposal.
	(d) Cleanliness when handling food.
	(e) Fly control.
Prevention of epidemics	If an isolated outbreak of typhoid occurs an intensive search for the source of the infection needs to be carried out. Man is the **reservoir of infection** for typhoid germs, that is, the germs depend on the human body for survival. They live and multiply there and are transmitted via food and water from person to person. Therefore the source of infection may be an infected person, or a carrier, or contaminated food or water. The source of infection must be treated as well as contacts with it to prevent or halt an epidemic.

WHOOPING-COUGH

Cause	Bacterium.
Transmission	Droplet infection or contact.
Incubation	7–10 days.
Symptoms	Spasmodic coughing.
Treatment	Medicine can relieve symptoms.

122

DISEASE

Prevention	(a) Immunisation of babies with dead whooping-cough bacilli. Three injections are given at monthly intervals with a booster dose at five years. This vaccine is part of the combined immunisation against whooping-cough, diphtheria and tetanus. Because of mass immunisation there has been a marked drop in the number of cases. The disease can be particularly severe in babies and infants.
	(b) Natural active immunity will be acquired by having the disease.

AMOEBIC DYSENTERY

Cause	Protozoan (*Entamoeba histolytica*).
Transmission	Contaminated water and food, especially raw vegetables.
Incubation	Usually 3–4 weeks.
Symptoms	Diarrhoea and other symptoms.
Treatment	Medicine to destroy the pathogens.
Prevention	(a) Hygienic methods of sewage disposal.
	(b) Cleanliness in preparing food.
	(c) Clean water supply.

MALARIA

Cause	Protozoan (*Plasmodium*—the malaria parasite, different species causing different forms of malaria).
Transmission	Mosquito (see Fig. 13.3).
Incubation	Usually 12–30 days.
Symptoms	Repeating cycle of chill, fever, sweating, interval.
Treatment	Paludrine and other drugs to destroy the parasite. Protection of the patient from mosquitoes to prevent the parasites from being transferred to another person.
Prevention	(a) An anti-malarial drug taken once or twice weekly when living in endemic areas.
	(b) Installation of screens to prevent the entry of mosquitoes into dwellings.
	(c) Sleeping under a mosquito net.
	(d) Spraying the inside of dwellings with an insecticide.
	(e) The use of insect repellents on uncovered skin.

In 1955 the World Health Organisation embarked on a worldwide campaign to eradicate the disease. The malaria parasite needs both man and the mosquito to complete its life-cycle (Fig. 13.3) and if either of them are infected the disease can spread. Therefore the campaign was based on:

1. Destroying the source of infection within the human body by using drugs to kill the parasite.

2. Destroying the means of transmission by killing the mosquito:
(a) By destroying the breeding grounds:
  (i) Draining swamps, marshes and stagnant pools.
  (ii) Turning slowly moving rivers into swiftly flowing water.

123

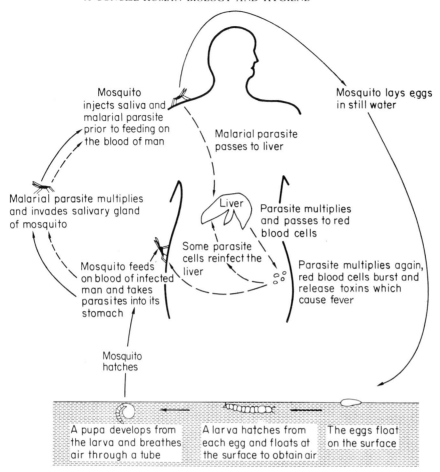

Mosquito
injects saliva and
malarial parasite
prior to feeding on
the blood of man

Mosquito lays eggs
in still water

Malarial parasite
passes to liver

Malarial parasite multiplies
and invades salivary gland
of mosquito

Liver

Parasite multiplies
and passes to red
blood cells

Some parasite
cells reinfect the
liver

Mosquito feeds
on blood of infected
man and takes
parasites into its
stomach

Parasite multiplies again,
red blood cells burst and
release toxins which
cause fever

Mosquito
hatches

A pupa develops from
the larva and breathes
air through a tube

A larva hatches from
each egg and floats at
the surface to obtain air

The eggs float
on the surface

(b) Spraying a film of oil containing insecticide onto lakes and ponds to destroy the mosquito larvae.

(c) Spraying the inside of dwellings and the surrounding neighbourhood with insecticides to destroy the adult mosquito.

Fig. 13.3 Diagram to show the life cycles of the mosquito (———>) and malarial parasite (— — >). The disease is difficult to eradicate from a man once he has been infected as only the stage occurring in the red blood cells can be controlled by drugs. The liver remains infected to some extent. Gradually, over a year or more this infection will decrease and the recurring bouts of fever will occur less frequently.

VENEREAL DISEASE

Venereal disease, usually referred to as V.D., is disease of the sex organs. The germs live in the sex organs and are transmitted from one person to another during sexual intercourse. As the germs die very quickly away from the heat of the body it is almost impossible to catch V.D. in any other way.

If the disease is treated early enough it can usually be cured, but it is a dangerous disease as the first symptoms are often painless and hidden, especially in women, and therefore many people do not realise they have V.D., until the infection has spread to other parts of the body. When this happens the effects can cause pain, severe illness and even death.

Most large hospitals have special V.D. clinics which give confidential advice and treatment. Anyone suspecting that they have V.D. should either consult the family doctor or, if preferred, visit the V.D. clinic for an examination. If V.D. is present, treatment will be given, and if it is not, then a cause for worry will have been removed. Brief details will be given of two forms of V.D., gonorrhoea, because it is the most common, and syphilis because it is extremely serious if left untreated.

GONORRHOEA

Cause	Bacterium.
Transmission	Personal contact during sexual intercourse.
Incubation & Symptoms	Men—2–10 days after infection, the penis becomes inflamed, with a yellow discharge appearing at the tip. A burning sensation is felt when passing urine. If not treated the germs will spread to inflame other parts of the body which may cause general ill-health, swollen joints and sterility.
	Women—over 50% show no early symptoms because the germs breed in the opening of the womb, the inflammation is not painful, and the yellow discharge goes unnoticed. Sometimes pain is felt when passing urine. If not treated, the germs will spread, causing inflammation and severe pain in other parts of the body, and possibly, if a woman with gonorrhoea has a baby, it may develop serious eye disease and could go blind.
Treatment	Antibiotics.
Prevention	No promiscuous intercourse.

SYPHILIS

Cause	Bacterium.
Transmission	Personal contact during sexual intercourse.
Incubation & Symptoms	3–6 weeks after infection. The first sign appears as a painless sore, usually on or near the sex organs. After days or weeks this disappears and the germs spread to all parts of the body. A few weeks later, the second stage occurs with fever, rash, sore throat or loss of hair. These symptoms also eventually disappear, and months or years later the third stage occurs. The disease will re-appear in one or more parts of the body, causing blindness, heart disease, deafness, insanity and death.
	A pregnant woman with untreated syphilis can pass the germ into the blood stream of her baby so that it is born dead or diseased.
Treatment	Antibiotics can cure the disease completely if treated in the first two stages, but the damage done to the tissues during the third stage cannot be repaired.
Prevention	No promiscuous intercourse.

OTHER COMMON PARASITES
Lice

Lice are small insects which either live on the body or amongst the clothing. They feed on blood which they suck from the skin and thus can be vectors (carriers) of disease, e.g. typhus.

125

Head Louse

This species lives amongst the hair. It lays small white eggs called nits which are attached by 'cement' to the base of the hairs. These take about a week to hatch and soon grow into adults and reproduce. This is a common parasite of young children, particularly those from poor, overcrowded areas, where it spreads from head to head.

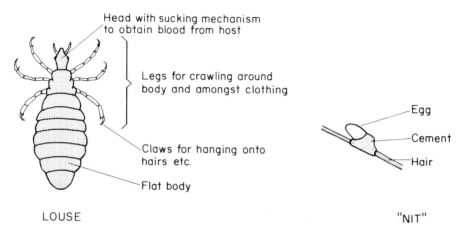

Head with sucking mechanism to obtain blood from host

Legs for crawling around body and amongst clothing

Claws for hanging onto hairs etc.

Flat body

LOUSE

Egg

Cement

Hair

"NIT"

Body Louse

This species lives and lays eggs in clothing and goes to suck blood from the body once or twice daily.

Pubic Louse

This species lives amongst the pubic hair and is more 'crab'-shaped.

Control   Adults are killed by various insecticides and eggs can be removed from the hair by a fine tooth-comb. Reinfestation can be prevented by frequent washing of body, hair and clothing.

Fleas

Fleas are small, wingless insects with legs adapted for jumping and they live in clothing next to the skin. The human flea does not carry disease but when it pierces the skin to suck blood, it leaves small, red marks which irritate. The rat flea, however, is responsible for transmitting bubonic plague from rats to man. This caused the Black Death of 1348 and the Great Plague of 1665. Fortunately the disease has been eradicated from this country.

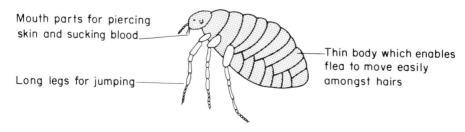

Mouth parts for piercing skin and sucking blood

Long legs for jumping

Thin body which enables flea to move easily amongst hairs

Control As they lay their eggs in crevices of buildings, furniture, bedding or wherever there is dirt, the most important control is cleanliness. Infested bedding needs to be burnt and the body and clothes washed frequently. Fleas are also killed by insecticides.

Bed Bug

This small insect lives in bedding and other cracks and crevices in the room. It comes out at night to feed by sucking human blood. This causes intense irritation of the skin and the bugs also give off a most unpleasant smell.

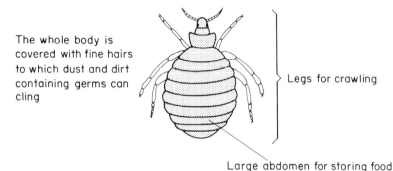

The whole body is covered with fine hairs to which dust and dirt containing germs can cling

Legs for crawling

Large abdomen for storing food

Control Fumigation and insecticides can kill them but they are difficult to eradicate.

Itch Mite

This is a very small animal with four pairs of legs. It burrows into the epidermis of the skin where it lives and lays eggs. This causes irritation, scratching and redness, especially between fingers and toes and at the wrist, and is known as **scabies.**

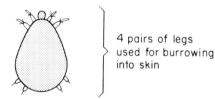

The itch mite has a simple body as most of its life is spent in tunnels in the epidermis

4 pairs of legs used for burrowing into skin

Control After a bath at night, apply benzyl benzoate emulsion to all parts of the body, except the eyes. Allow to dry, retire to bed, bath the following morning. This should kill the itch mite in the skin. Change underclothing and sheets, and wash to kill mites and their eggs contained in the epidermis which has rubbed off. Also iron heavy clothing, especially around cuffs for the same reason.

Ringworm

Ringworm is caused by a fungus which infects the skin and grows outwards from the point of infection to form a ring. It causes irritation and scratching and is very easily passed on from one person to another. **Athlete's Foot** is ringworm of the feet, causing the skin between the toes to turn white and peel off. It is easily transmitted in public swimming pools.

Control A fungicide which is effective against ringworm will clear the skin of this infection.

Threadworms

Threadworms are white, threadlike worms, about $\frac{1}{3}$–1 cm long, which live in the large intestine. They are found quite commonly in young children and will be seen in the faeces. They cause irritation around the anus as the females crawl out at night to lay their eggs. This encourages the child to scratch and hence to distribute the eggs. These can be picked up by the same child, or a different one, when they put their fingers or toys in the mouth. The eggs pass to the intestine where they hatch.

Control Medicine to clear the worms from the intestine. General cleanliness.

Tapeworms

Tapeworms are parasites which live in the intestines and feed on the digested food passing through. Each is like a long piece of white tape with suckers and hooks on the head for attachment to the

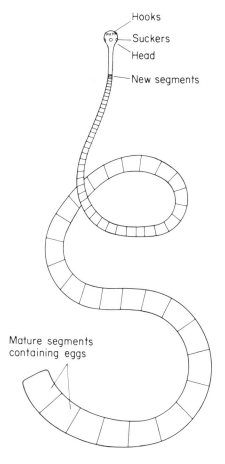

Hooks
Suckers
Head
New segments

Mature segments containing eggs

Fig. 13.4 The tapeworm.

intestine wall (Fig. 13.4). The head gives rise to a long chain of segments which gradually increase in size as they mature. New segments are continuously being formed, and at the other end of the tapeworm, the older segments break off to be expelled with the faeces. These segments are filled with eggs which do not develop unless they are eaten by a second host, either a pig in the case of the pork tapeworm, or cattle in the case of the beef tapeworm. In the second host the eggs hatch into larvae which burrow into muscle tissue and form cysts. New tapeworms can only develop from the cysts if man eats uncooked pork or beef containing them.

Tapeworms are examples of parasites which need two hosts to complete the life cycle. Should the eggs of the pork tapeworm be eaten by man these can form cysts, but not tapeworms. These cysts may cause grave damage if they develop in heart, eye or nervous tissue.

Control (a) Medicine to dislodge the head of the tapeworm so that it can be expelled with the faeces. (b) The inspection of meat for tapeworm cysts. (c) The thorough cooking of suspected meat to destroy the cysts.

RABIES

Cause	Virus.
Transmission	Saliva of a rabid animal (one suffering from rabies).
Incubation	Varies, but usually 2–3 months.
Symptoms	Fever, delirium, convulsions, paralysis. As the disease develops the patient alternates between furious convulsions and periods of calm. The throat muscles tighten so that it is impossible even to drink fluids; hence the name **Hydrophobia**—fear of water.
Treatment	None; the patient dies within a few days.
Prevention if bitten	If bitten, scratched or even licked by a rabid animal, (a) Immediately wash the area with water and soap, or detergent or disinfectant to remove any virus before it can attach itself to nervous tissue. (b) Rabies vaccine to make the body produce antibodies. (c) Rabies serum for immediate immunity.
Protection of the Community	Quarantine restrictions to prevent dogs, cats and other animals from bringing the disease into a rabies-free area.

This disease causes a terrifying death and the patient can remain aware of what is happening right up to the end. The virus attacks the nervous system and, when it enters the skin, travels slowly along the nerves to the C.N.S. It is also able to multiply in other organs such as the salivary glands. Although rabies can infect all warm-blooded animals—foxes, bats, jackals, etc., the disease only becomes a serious risk to people if domestic animals are infected. This is why there is a need for the quarantine of dogs and cats. Rabies in dogs (and cats) takes two forms—furious rabies and dumb rabies. In **furious rabies** the dog produces much saliva and will bite anything in its fits of rage (i.e. the mad dog). In **dumb rabies** the dog simply becomes apathetic and depressed and the disease may not be diagnosed (although the saliva is still highly infectious).

Chapter Fourteen

HYGIENE

Hygiene has two meanings. In popular usage it means cleanliness, both personal and environmental, for the maintenance of health. However, it also has a wider definition meaning a study of the principles governing health and a person whose life is lived according to these principles will be far less prone to disease. Therefore hygiene includes the study of:

Personal and environmental cleanliness.
A balanced diet and regular meals.
Exercise and fatigue.
Rest and sleep.
Posture.
Suitable clothing.
The importance of fresh air.
The need for self-discipline and the dangers of bad habits.

We shall consider all these topics in this chapter except environmental cleanliness which will be dealt with in Chapter 15.

PERSONAL CLEANLINESS

Personal cleanliness is essential for the health of the individual and for the health of the community. It is necessary to:

1. Prevent germs from gaining entry to the body and causing disease. If they do, effort must be made to prevent them from being passed on to others.

2. Deter parasites such as lice and fleas from living on the body or in the clothing.

3. Prevent the body from developing a smell objectionable to other people.

Cleanliness of the Skin

The skin can be kept clean by frequent washing with soap. Besides removing dirt this also gets rid of unwanted epidermal scales, sebum, and stale sweat. These all provide breeding grounds for germs, and it is the bacteria that thrive in stale sweat that give rise to the unpleasant smell called 'body odour'. Washing also helps to prevent the pores of the skin from becoming blocked, thus preventing them from functioning properly. Blocked pores often lead to the formation of unsightly blackheads.

Hands should be washed and nails scrubbed before preparing or eating food as this helps prevent the spread of a number of infectious diseases. It is also most important that hands are washed after visiting the lavatory as the faeces contain large numbers of bacteria which are able to penetrate through paper and may be a source of disease.

Cleanliness of the Hair

Washing the hair not only removes dirt and grease but also discourages the head louse (p. 126). Some people produce more grease (sebum) than others and need to wash their hair more frequently.

Cleanliness of the Teeth

These should be cleaned after meals to remove food particles from around the teeth. The food provides a breeding ground for bacteria. Some bacteria can produce acids which dissolve the enamel causing tooth decay, pain, and sometimes infection of the gums. Their activity also causes an unpleasant smell which is one of the causes of 'bad breath'.

Cleanliness of Clothing

Clothing worn next to the skin needs to be washed frequently to remove epidermal scales, sweat, urine, faeces, and other dirt. All these provide breeding grounds for germs which can then be transmitted from one person to another. Other clothing needs to be kept clean by washing or dry-cleaning to discourage fleas, lice, etc. Dirty clothing also produces an unpleasant smell.

BALANCED DIET AND REGULAR MEALS

A balanced diet is essential in order to provide all the food substances necessary for life, and in the right proportions (p. 102), whereas too much carbohydrate and fat may cause obesity (p. 134).

Regular meals help to keep the digestive system in good working order. An unbalanced diet and irregular meals can both cause indigestion and other digestive troubles.

THE IMPORTANCE OF EXERCISE

Exercise involves movement, and regular use of all the various parts of the body keeps it in good working order for the following reasons:

1. Muscles need to be used to prevent atrophy (wasting away), and regular exercise can strengthen muscles.

2. **Muscle tone** is improved. Muscle tone is the slight state of tension in which all skeletal muscles are kept, and this prevents flabbiness and enables them to respond more quickly and easily to demands made upon them.

3. Peristalsis is increased which helps to prevent constipation.

4. Breathing becomes deeper allowing full use to be made of the lungs.

5. Regular exercise trains the nervous system to improve coordination of body movements.

6. Exercise uses up food and this can prevent the accumulation of too much fat in the body.

7. After exercise the body needs to relax and this can be an aid to sound sleep.

For some people (e.g. manual workers, active housewives) their job provides all the exercise they need. Others (e.g. office workers) may need to take regular exercise to keep fit, such as walking, swimming, dancing, gardening.

Vigorous exercise should not be undertaken without training otherwise it might strain the muscles and cause them to ache.

Vigorous exercise should not be taken after a heavy meal because a large proportion of the blood is diverted to the stomach and intestines and is therefore not available to the skeletal muscles. Hence it is unwise to swim soon after a meal as the muscles of the limbs may suffer from cramp.

Excessive exercise will cause fatigue.

FATIGUE

Fatigue is the state of tiredness and can be due to:

1. Muscle tiredness. When a muscle is working it produces lactic acid and this will accumulate unless sufficient oxygen is supplied to the tissue to oxidise it. As lactic acid accumulates it causes pain to the muscle and impedes its action. The muscle then needs to rest until the acid has been removed by oxidation.

2. Tiredness of the central nervous system. This results in lack of co-ordination of muscles. The body then needs to rest, and in this way is prevented from over-exertion and exhaustion.

3. Psychological causes such as boredom, emotional strain, a repetitive task or daily routine.

Exhaustion is extreme fatigue.

REST

Rest is necessary when the body needs to stop using the tissues which have been working in order to allow them to recover. The work can either be physical or mental, and rest can be taken in the form of relaxation or sleep or change of occupation.

After the whole body has been working it is necessary to sit or lie down to allow:

the muscles to relax and get rid of lactic acid.

the rate of heart beat to fall.

the temperature of the body to drop.

the nervous system to rest.

If only part of the body has been working then a change of activity will allow that part to rest.

A clerk can rest his brain by gardening. A gardener can rest his muscles by stopping to read the newspaper.

SLEEP

Sleep is a form of unconsciousness from which it is possible to awaken with relative ease. The body needs sleep in order that the nervous system can function efficiently. Lack of sleep causes irritability, lack of concentration and unco-ordinated body movements. Severe deprivation can cause hallucinations and delusions.

During sleep:

The metabolic rate falls.

The rate of heart beat is reduced.

Breathing is shallow.

Less urine is produced.

The sense organs and nervous system are active only to a limited extent, but noise can wake the person up, the skin is sensitive to touch, and dreaming occurs.

The amount of sleep needed varies with the person but it is related to age, babies needing more than adolescents, who need more than older people.

Good sound sleep can be encouraged by:
Physical exercise which tires the body.
A warm, comfortable bed.
A quiet, darkened room.
The ability to put worries aside.

POSTURE

Posture is the way in which the parts of the skeleton are held in position in relation to each other. The skeletal muscles are responsible for posture and it is desirable that they should have good muscle tone, that is, that they should be in a partial state of tension.

Good Posture

Good posture is one in which the bones are held in the most favourable position for the maintenance of good health, and with minimum effort on the part of the skeletal muscles.

A good standing posture is when the body is held erect with the shoulders well back and the weight evenly distributed on both feet. The abdominal muscles should be in a slight state of contraction and the arms hang loosely at the sides.

A good sitting posture is when the pelvis and back are supported by the chair, and the feet are able to be placed comfortably on the floor.

Effects of Bad Posture

1. The muscles of the back and abdomen are strained, causing pain.
2. When the shoulders are rounded, breathing is impaired.
3. It detracts from the appearance.
4. It is habit-forming, therefore it is particularly important that children should acquire good posture.

SUITABLE CLOTHING

Clothing is necessary because, unlike other warm-blooded animals, man does not have fur or feathers to keep him warm. Suitable clothing is that which enables the person to keep the body temperature constant at about $37\,^{\circ}C$ ($98\cdot4\,^{\circ}F$). The amount and type of clothing needed will depend on the activity or inactivity of the body, and the temperature of the environment.

If the person is taking exercise or digesting a meal, heat energy will be generated by the muscles and circulated around the body by the blood. Therefore less clothing is necessary than when the person is inactive, when much less heat will be generated.

If the temperature of the air is cold the body needs to be well covered with materials like wool in order to trap a layer of air around itself. This will act as an insulating layer preventing both the outside cold from penetrating through to the skin and the warmth generated by the body from being lost.

133

If the surrounding temperature is hot the body needs much less clothing so that excess heat can be lost easily; this prevents the body from over-heating. But there must be enough clothing to prevent sunburn and it should be made of material like cotton, that does not retain the heat and can absorb sweat.

Although the main function of clothing is to keep the body warm it should also:

1. Not restrict movement by being too tight, e.g. trousers in which it is impossible to sit down.

2. Not inhibit circulation, e.g. garters that restrict the blood supply to the legs.

3. Be shower-proof in rain to prevent the body from becoming chilled as this can lower resistance to infection. Waterproof clothing should not be worn for any length of time as it prevents evaporation of sweat, causing discomfort.

4. Include shoes which are comfortable as this encourages good posture and prevents trouble with the feet. It is particularly important that children should have shoes which are large enough so that the feet can grow normally. If the feet become deformed, foot trouble may occur in later life.

IMPORTANCE OF FRESH AIR

When a person breathes in fresh air it contains an abundance of oxygen which has been released from plants during photosynthesis. This has an invigorating effect as the extra oxygen taken up by the body when breathing allows increased energy production. Fresh air also contains fewer germs as many will have been killed by sunlight or dehydration.

The air in an unventilated room containing people quickly becomes unhealthy and is said to be 'stuffy'. This is because:

1. The carbon dioxide content rises and the oxygen content drops thus making respiration more difficult and causing drowsiness.

2. The water vapour given off by the lungs causes the moisture content of the air to rise. Discomfort arises when sweat is unable to evaporate.

3. Heat from the bodies will cause the room temperature to rise which may cause discomfort.

4. The risk of infection increases as airborne germs can easily be spread from one person to another.

SELF-DISCIPLINE

It is a general rule that anything taken to excess is harmful whether it is work, play, food, alcohol etc. Strain will be put upon one or more systems of the body thus rendering it more prone to disease.

Self-discipline needs to be exercised to maintain a state of health by preventing over-indulgence in any one aspect of living. We shall consider four aspects of living that can lead to habit formation, and as they harm the health they can be considered bad habits.

Dangers of Obesity

Appetite is no guide to the amount and type of food required by the body. The habitual eating of too much food may cause excessive fat to be deposited under the skin and around the internal organs. The result is **obesity** (fatness). The effects of obesity on the body are:

1. It is the commonest cause of breathing difficulties.

2. The skeleton has more weight to support which puts greater strain on the joints, especially the hip joint.

3. It is more difficult to take exercise and muscles become flabby.

4. High blood pressure may result.

5. Varicose veins are more common.

6. Expectancy of life is shorter.

Dieting to Reduce Weight

It is inadvisable to attempt to reduce weight too quickly, and not at a greater rate than 3–4 lbs (or about 1·5 kg) weekly, unless under medical direction. The intake of fats, carbohydrates and alcohol should be reduced as much as possible, but a sufficient supply of protein, fruit and green vegetables should be maintained. Exercise also helps to reduce weight.

The Dangers of Alcohol

Alcohol is produced by the fermentation of yeast and throughout man's history alcoholic drinks have been made.

Taken in strict moderation alcohol acts as a stimulant which does little harm and can be beneficial. It relieves nervous tension, makes conversation easier, dulls mild pain, has a warming effect and stimulates the appetite.

But alcohol taken in excess is a poison. It is quickly absorbed into the body and assimilated by the tissues, the absorption being more rapid when the stomach is empty. Some tissues are more affected by alcohol than others and it continues to have an effect until the body has had time to destroy it by oxidation.

Effect on the Nervous System

Alcohol is a **depressant** as it has an anaesthetic effect on the nervous system.

The first effects are to dull mild pain and relieve worries and anxieties. As alcohol reaches the cortex of the brain, will-power and self-control are reduced and inhibitions are released. This gives rise to a feeling of gaiety and the mistaken impression that alcohol is a stimulant. The emotions are affected when alcohol reaches the front part of the brain and may result in joy and laughter or sadness and weeping. The vision and speech areas are the next to be affected resulting in blurred vision and slurred speech. When alcohol is absorbed by the cerebellum, muscle movements become unco-ordinated causing walking difficulties. Finally when the brain is completely anaesthetised the person becomes unconscious.

Too much alcohol is dangerous because:

1. It reduces the person's self-control. In this state the person may behave in a manner that is regretted after the effects of the alcohol have worn off.

2. It reduces the person's ability to do jobs which require skill and judgement, like driving a car, and this is dangerous both for the driver and other people.

3. After drinking an excessive amount the brain is completely anaesthetised and the person falls into a drunken sleep until the alcohol absorbed by the nervous system has been destroyed.

4. If taken frequently it may lead to chronic **alcoholism** which is a disease, the patient being addicted to alcohol. This may be accompanied by digestive disorders, cirrhosis of the liver (destruction of liver cells), heart trouble, hallucinations and other nervous disorders, etc. Treatment in hospital can cure alcoholism and this has to be followed by complete abstinence. **Alcoholics Anonymous** is an informal organisation of men and women for whom alcohol is a problem.

The Dangers of Smoking

There is sufficient medical evidence to show that cigarette smoking can be a major cause of ill-health. Smoking is habit-forming and some people find that it is both pleasant and soothing. But it is also accompanied by dangers and it is a significant fact that many doctors have given up smoking altogether.

Tobacco smoke is a complex mixture of gases and minute droplets of tar. The composition of tobacco smoke varies with the type of tobacco, the method of curing and the way it is smoked. Nearly 1000 different substances have been identified in it including:

1. Cancer-producing substances that cause cancer in experimental animals.

2. Irritant substances that irritate the bronchial tubes and may be responsible for coughing and bronchitis.

3. Nicotine This has adverse effects on the heart and blood vessels. It also affects the nervous system and is probably the substance responsible for a smoker becoming dependent on tobacco.

4. Carbon monoxide which interferes with the capacity of the blood to carry oxygen to the tissues of the body.

The effects of tobacco smoke on the body are extremely complicated and still require a great deal of research. However, certain facts have been established:

1. The greater the number of cigarettes smoked per day the greater the health risk.

2. The younger the person is when smoking begins the greater the damage is likely to be.

3. Smokers who inhale deeply run greater health risks than those who don't.

4. Heavy smokers have a shorter life expectancy because they are more prone to destruction of lung tissue and chronic bronchitis, heart disease and lung cancer.

5. Mothers who smoke during pregnancy tend to have smaller babies than non-smokers and may be more likely to lose their babies by miscarriages, still-births or deaths in the first few days after birth.

6. Smoke irritates the lining of the lungs. This produces mucus, and the result is 'smoker's cough'.

7. Smoke can be a great annoyance to non-smokers if it causes them to suffer more from bronchitis, asthma or other lung complaints.

8. When a smoker gives up smoking the risk to health gradually decreases, and after ten years has almost disappeared.

136

The Dangers of Drug-taking

A drug is a chemical substance which has an effect on the body. Many drugs, when taken under medical supervision, can be used to prevent or fight disease, and in this connection they are therefore beneficial to man. However, the term 'drug-taking' refers to the misuse of a few drugs which affect the nervous system. There are three groups:

1. Drugs which alter the senses. For example, **cannabis,** also called marijuana, pot, weed, grass and many other names, creates an illusion of well-being. **LSD (lysergic acid)** affects the eyes and brain and produces hallucinations. The mind feels as though it is going on a journey or 'trip', and this may cause the person to do strange things.

2. Drugs which stimulate the nervous system, causing it to speed up the rate at which the body works. Examples of such drugs are the **amphetamines** and other pep pills which produce a feeling of boundless energy. This is followed by depression when the effects of the drug wear off, which, in turn, leads to the desire for another pep pill.

3. Drugs which act as depressants, as they have an anaesthetic effect upon the nervous system. **Barbiturates** are used as sleeping pills. However, their continual use produces a state in which sleep is impossible without them. **Narcotics*** cause narcosis, a trance-like state just short of sleep. Included amongst the narcotics are **cocaine, opium** and the opium derivatives—**morphine** and **heroin.**

Drug-taking is dangerous for the following reasons:

1. The effect that any particular drug will have cannot be predicted as it will vary according to the emotional state of the person concerned, and will also have different effects on different people.

2. Because these drugs affect the nervous system, they can cause a change in personality. This can show itself as severe depression, irresponsibility, slovenliness, violence or other undesirable traits.

3. When the senses are affected, this may produce a sense of unreality (for example, the person may attempt to fly) or result in hallucinations in which things are seen and heard which are not really there. This can be terrifying and have disastrous results.

4. Although at first a small quantity of the drug will cause an effect, it will need to be taken in ever increasing doses to cause an acceptable effect. The person is then more likely to be dependent on the drug.

5. Drug dependence can lead to drug addiction.

Drug-dependence and Drug-addiction

Drugs of dependence, the 'soft' drugs, are those which cause a person to have a psychological need for the effect that the drug produces. For example, sleep from barbiturates, energy from amphetamines or a sense of well-being from cannabis. Nicotine is also one of the 'soft' drugs.

These drugs do not cause any physical change in the body and are generally regarded as 'habit-forming' but non-addictive. However, many of those who become dependent on such drugs find

* In U.S.A. the term 'Narcotics' is applied to all the banned drugs.

it difficult to give them up once they have become conditioned to their use, and may suffer severe mental and physical stress when a 'soft' drug is taken away from them.

A dependence on 'soft' drugs may lead to the taking of **drugs of addiction,** the 'hard' drugs, although there are many cases in which people have become addicted to 'hard' drugs without previous recourse to 'soft' drugs. Addictive drugs cause physical changes in the body, and the majority of users find it extremely difficult to give them up. The most common examples in use today are alcohol, morphine and heroin. An addict cut off from his supply of these drugs experiences withdrawal symptoms, which can be very painful. An addict who wishes to give up a 'hard' drug will usually require assistance, which can be given most effectively in special clinics.

The taking of drugs creates several social and health problems for the community. In many countries it is illegal to possess or to use drugs, and hence drugs are sold in the black market. Pedlars ('pushers') can charge very high prices and are always on the lookout for new people to introduce to drugs. Once a person has become dependent on or addicted to a drug they will have to keep going back to the pedlar for their supply. If a supply is not forthcoming or the addict does not have sufficient money to buy the drugs, he may resort to stealing to obtain what he needs. Much crime centres round the use of drugs, especially in drug-trafficking and selling, which can involve large sums of money.

Drug-taking may start in a variety of different ways. Usually it is the weaker-minded people who become involved. It may start as an experiment, which is found to be pleasurable to begin with; it may be a means of escape from personal problems; it may merely become the thing to do amongst a group of people who are not fully aware of the dangers involved. A person who graduates to heroin from one of the soft drugs soon begins to find that he has to take more and more to satisfy his craving. He then loses interest in his family and friends and has no desire for sex or any pleasure other than the drug. His health rapidly deteriorates and if the addiction continues the result is a wretched and early death.

RESPONSIBILITIES OF PARENTS FOR THE HEALTH AND TRAINING OF THEIR CHILDREN

Children are completely dependent on their parents or guardians for their health and well-being.

It is therefore the duty of these adults to:

1. Ensure the children have a balanced diet so that their bodies can grow and develop normally, and that deficiency diseases are prevented.

2. Keep them clean to prevent germs from causing ill-health as far as possible. It is also essential to train the children to keep themselves clean.

3. See that they have adequate medical attention when ill.

4. Clothe them adequately to keep them warm.

5. Make sure they clean their teeth daily and take them to visit the dentist at regular intervals. They will then not need false teeth at an early age.

6. Allow them to have plenty of exercise and encourage good posture, as it is during childhood that the bones develop.

7. Make sure that they develop good habits and not bad ones. Habits are conditioned reflexes and good habits developed during childhood are likely to be continued in adult life. On the other hand, a bad habit is difficult to break as it means unlearning a conditioned reflex.

Chapter Fifteen

GROWTH AND DEVELOPMENT

Growth is an increase in size.
Development is an increase in complexity.

FACTORS WHICH INFLUENCE GROWTH
AND DEVELOPMENT
1. **Genetic factors (inherited factors) provide a blue-print for growth and development** The genes which a child inherits from his (her) parents
(a) govern
cell activity and the way in which cells are arranged into tissues and organs to produce a body with the shape of a human being,
the sex of the child,
skin colour, hair type, shape of nose, blood group, rhesus factor, and many others.
(b) influence
the way in which the body functions,
the rate of growth and development, for example, the age at which the teeth appear and when the child is ready to sit up, walk and talk,
the inborn level of intelligence,
the inborn temperament, particularly by control of hormone output. They have an effect on the child's temperament and behaviour, resulting in one who is by nature placid, lively, nervous, sleepy, affectionate, aggressive, and so on.
2. **Environmental factors interact with inherited factors**
Environmental factors include:
the housing conditions under which the child lives,
who looks after him,
whether he is loved and wanted,
how he is fed and generally cared for,
the sort of companionship he has,
whether he is encouraged to learn, or is ignored, or prevented from learning by over-protection.

INTERACTION OF GENETIC AND ENVIRONMENTAL FACTORS

Growth and development in the womb are mainly under genetic control and follow a pre-determined path. At birth, babies may seem very much alike. But by the time they have become adults, they show tremendous variation due to the interaction of genetic and environmental factors at every successive stage of their growth and development.

Environmental factors affect a child's inborn nature either by

reinforcing inherited characteristics, modifying them, or counter-
acting them, for example:

a nervous child may be taught to become confident,

a placid child may acquire the habits of aggression,

prolonged malnutrition may prevent a child's genes for height
from having their full effect, or may affect brain development
and consequently the child's intelligence.

INDIVIDUALITY AND PERSONALITY

Each person develops into a recognisable individual with his
own appearance, personality and way of behaving because:

1. people are genetically different,

2. environmental factors vary for each child.

The complex interaction between heredity and environment
makes individual development very hard to predict. Even if it were
possible to rear all children in an equal environment they would
grow up to look different and develop differing abilities, tastes and
interests. This is so for identical twins, even though they have the
same genes; they develop into individuals who do not exactly
physically resemble one another and who each have a distinct
personality. This is probably due to differing environmental factors,
for example, they lie in different positions in the womb, one might
have a birth mark, and so on.

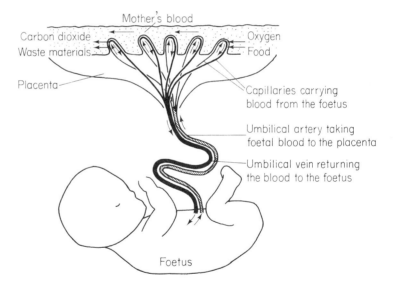

Fig. 15.1 Diagram of placenta.

LIFE IN THE WOMB (see also pp 87–8)

While the baby is in the womb it is in a little protective pond
with the water (amniotic fluid) kept at 37°C. The mother breathes
for the foetus, digests its food, excretes its waste, maintains its
body at the right temperature and protects it against disease. The
foetus grows and its heart beats, circulating foetal blood around
its body and along the umbilical cord to the placenta and back
again (Fig. 15.1).

The placenta is the link between mother and child. Basically it is a much-folded membrane with the mother's blood on one side and the foetal blood on the other. Substances which are in small enough particles are able to pass across the placenta; food substances and oxygen diffuse from the mother's blood to the foetal blood, and carbon dioxide and other waste substances diffuse in the opposite direction.

Viruses, alcohol, antibodies, chemicals from smoke and from medicines taken by the mother are also able to diffuse across the placenta. Some of these substances may affect the foetus; others do not; in some cases it depends on the stage of development that has been reached (pp 143–5).

Growth and Development in the Womb

By the end of the second month the main structures of the body are more or less in place and the heart is beating; the embryo now shows human form and is referred to as a foetus. During the third month the nerves and muscles develop rapidly and by the end of the third month, the foetus can swallow, frown, make a fist, and move by turning his head and kicking, although the mother does not feel these movements until about the fifth month.

Whilst in the womb, the baby's body is covered by **vernix**—a greasy, whitish, protective substance. There is also a covering of fine hairs—**lanugo**—which is usually shed before birth.

If the foetus leaves the womb before the 28th week it is known as a miscarriage; after the 28th week it is regarded as a birth. Most babies born at 28 weeks do not survive and occasionally one born younger than 28 weeks does.

ANTENATAL CARE (ante = before, natal = birth)

The care of the baby while it is in the womb is important for its future development.

Antenatal clinics monitor the health of the pregnant mother and the progress of the foetus, to safeguard the mother's health and to ensure that the baby has the best possible start in life.

Tests that are carried out include:

Blood test A small sample of blood is taken from the mother to test for: Rh factor (p 25).

Anaemia. The commonest form of anaemia in pregnant women is iron-deficiency anaemia as the foetus takes what iron it needs at the expense of the mother.

Blood pressure A careful check is kept on the mother's blood pressure as a rise may indicate 'toxaemia of pregnancy'—a condition which needs immediate treatment.

Urine test to detect:

glucose, as it may indicate diabetes,

protein, as it may indicate renal disease of some kind.

Weight check On average a pregnant woman gains about 450 g (1 lb) a week. An abnormal weight gain is likely to mean that she is eating too much and getting fat, but it may indicate that she is retaining too much water.

Chest X-ray to detect pulmonary tuberculosis, which fortunately is now uncommon.

These days X-rays are rarely used on the abdomen of a pregnant woman as there is the slight possibility that they may encourage the development of leukaemia in the baby, or damage the egg cells in the mother's ovary.

Ultrasound is used to obtain information from the womb about the position of the foetus, or its size, or to detect the presence of twins. This technique uses echo sounds to produce a pattern on a screen, and there is no evidence that it harms the foetus in any way.

Foetal heart rate   The baby's heart can be heard in the second half of pregnancy through a stethoscope placed on the mother's abdomen. It beats much faster than the mother's heart.

Amniocentesis   In some cases amniocentesis is carried out, that is, testing the amniotic fluid. A sample of amniotic fluid can be obtained by pushing a needle through the abdominal wall and into the womb.

Analysis of the fluid reveals certain diseases such as:

spina bifida (defect of the spine)

Rh factor incompatibility.

Cells from the surface of the foetus are constantly being shed into the amniotic fluid and they remain alive for some time. Live cells from the fluid can be cultured in a laboratory, that is, put into conditions in which they can grow and divide; their chromosomes become visible during cell division, and by examining them it is possible to detect, for example:

Down's syndrome (p. 144)

the sex of the child

the single gene responsible for thalassaemia—a type of anaemia.

MATERNAL FACTORS KNOWN TO AFFECT THE FOETUS

Nutrition

During pregnancy the well-nourished mother does not need to eat much more than usual, but the quality of food is important; her diet needs to have enough protein, vitamins and minerals for both herself and the baby. The baby will take what food it needs and the mother will suffer if she eats too much or too little.

The situation is different for mothers who suffer long-term (chronic) undernourishment; they produce underweight babies, and it is thought that vitamin deficiency may lead to malformation of the foetus. There is also a risk that the baby will be mentally deficient as malnutrition affects brain development during the last weeks of pregnancy and the first three years of life when the brain is growing and developing at its fastest.

German Measles (Rubella)

If the mother has German measles during the first four months of pregnancy the virus may seriously affect the child. It passes across the placenta and into the developing embryo where it multiplies freely; if the organs are at a certain stage of growth it causes them to develop in an abnormal manner. This may kill the foetus and result in miscarriage or, if the child survives, it may be deaf, blind, mentally retarded or have heart disease.

German measles is a mild virus disease and one attack usually

143

gives protection for life. It is spread by droplet infection and takes about 14–21 days to incubate. The symptoms are enlarged glands, rash and sometimes mild fever, and the patients should be kept away from mothers who are in the early stages of pregnancy, especially if they have not had the disease or been vaccinated against it.

Prevention   German measles vaccine is offered to all girls between the ages of 11–13; one injection is sufficient and usually gives immunity for life. If a mother in the early stages of pregnancy is exposed to a case of German measles she may be given an injection of antibodies (gamma-globulin) if she has not either had the disease or been vaccinated against it in order to give immediate, but temporary, immunity.

Age

With older mothers there is a greater risk of their babies being born with a congenital abnormality, that is, imperfect development of the baby while it is in the womb.

The commonest abnormality is **Down's Syndrome** (mongolism). When a mother is in her twenties the risk of having a Down's baby is 1 in 1500, this is increased to 1 in 270 in her late thirties and to 1 in 16 over the age of 45. It seems that as the ovary gets older it may become less efficient and produce eggs which have an extra chromosome—24 instead of the normal 23. If fertilised, the egg then contains 47 chromosomes and the extra chromosome influences both mental and physical development. The result is a Down's baby with a mental handicap and a distinctive appearance, such as slanting eyes. The degree to which a mongol child is affected varies, and they are usually friendly, cheerful and active, and loving care and training can encourage their mental development. A chromosome test on the cells in the amniotic fluid can detect Down's syndrome in the unborn child because of the presence of the extra chromosome.

Illness of the Mother

Serious illness during pregnancy does not necessarily affect the foetus; on the other hand a mild infection like German measles can cause serious damage. It is probably wise, if possible, to avoid infection during pregnancy because it is not known whether or not certain infections can harm the foetus.

Medicines

Mothers commonly take medicines of one sort or another during pregnancy and no harm comes to the child. But as a safety precaution it is wise to check that any medicines taken at this time are safe for the developing child. A few drugs are known to have an effect of which thalidomide is the most notable example, but there are others such as some anti-malarial drugs.

Thalidomide is a sedative drug which was taken to reduce the sickness that some mothers suffer during the early months of pregnancy. It is now known that this drug affects the embryo in the early weeks of development when the limbs are taking shape.

Thalidomide affects the growth of nerve cells in the embryo and if the nerves of a limb fail to develop so do the bones and muscles. Because of the thalidomide tragedy, new drugs are screened more thoroughly than before to discover any side-effects.

Alcohol

There is no evidence that alcohol in moderation has any effect on the developing foetus.

SMOKING

Evidence that Smoking has an Effect on Pregnancy

1. Mothers who smoke produce, on average, babies which are 200 g (almost $\frac{1}{2}$ lb) lighter at birth.

2. The lower birth weight is due to a slower rate of growth rather than a shorter pregnancy.

3. The effect of smoking is greater during the latter part of pregnancy. Mothers who give up smoking by the fourth month produce babies with birthweights similar to those whose mothers had never smoked.

4. Babies weighing less than $2\frac{1}{2}$ kg ($5\frac{1}{2}$ lb) at birth were nearly twice as common amongst mothers who smoked.

5. Stillbirth or death in the first week of life occurs nearly 30% more often in babies whose mothers smoked after the fourth month of pregnancy.

Smoking and Growth

The way in which smoking retards growth is not known. Direct research is almost impossible as controlled experiments on pregnant women are not feasible, and conclusions drawn from research on other animals may not apply to humans. What is certain is that when a mother smokes, various chemical substances pass from the smoke in the lungs into her bloodstream and will inevitably be carried to the placenta where they diffuse into the bloodstream of the foetus. Harmful substances in smoke which might affect the foetus are:

Nicotine

(i) by having a direct effect on growing tissue.
(ii) by causing narrowing of the blood vessels, thus restricting the amount of food and oxygen reaching the foetus.

Carbon monoxide

This becomes more concentrated in the blood of the foetus than in the mother's blood (because the foetus has a slightly different type of haemoglobin) and diminishes the amount of oxygen that the red cells can carry. It has been shown that the greater the amount of carbon monoxide in the baby's blood the lower the birth weight tends to be.

BIRTHWEIGHT

The average birthweight in the UK is about $3\frac{1}{2}$ kg ($7\frac{1}{2}$ lb) for a boy and slightly less for a girl.

Birthweight lower than average has three causes:—

(i) the baby inherits his small size, probably having small parents.

(ii) the baby is born prematurely and has not had enough time in the womb to grow to its full size.

(iii) the baby grows at a slower rate; this may be due to:
lack of food because the mother is severely undernourished; smoking.

(iv) toxaemia of pregnancy, a condition which reduces the blood supply to the foetus, thus starving the child.

Babies in category (i) are perfectly normal and no problem. The low birthweight of other babies is often due to a combination of circumstances; they are weaker and less able to cope with the strenuous business of being born and of immediately living an independent existence outside the womb.

BIRTH (The birth process is discussed on p. 88)

At the moment of birth so much happens so quickly. The baby emerges from the womb and changes rapidly take place to enable the child to live a separate existence from its mother.

1. It can stretch out and move unhindered.

2. It has to get its own oxygen through its lungs instead of via the placenta, so it has to start breathing almost immediately.

3. Blood has to be diverted from the placenta to the lungs to collect oxygen.

4. It must start to digest its own food, so the digestive system begins to work within a day.

5. It has to get rid of its own waste products, so the excretory system begins to work within a day.

6. The sensory nerve endings respond to light, sound, touch and pain.

7. The nervous system carries out reflex actions such as sucking, swallowing, blinking and withdrawing from painful stimuli.

8. The baby has to be able to deal with the germs to which it becomes exposed. It is helped in this by antibodies obtained from the mother through the placenta before birth, and afterwards by antibodies in colostrum and breast milk (pp 88–9).

Changes in the Baby's Blood System

When the baby emerges from the uterus it is still attached by the umbilical cord to the placenta, but:

1. Strong contractions in the uterus cut off the blood supply to the placenta and consequently the oxygen supply to the baby gradually runs short.

2. The placenta starts to separate from the uterus.

3. The umbilical artery to the placenta contracts, thus preventing the baby's blood from flowing to the placenta, but blood continues to flow through the umbilical vein until most of it has returned to the baby. When pulsating movements in the cord have ceased it is cut and tied. (Fig. 15.2).

4. The **ductus arteriosus** contracts and closes.

5. The valve—**foramen ovale**—between the right and left atrium closes.

6. So all the blood has to circulate through the lungs.

Fig. 15.2 Diagram of foetal circulation.
Notes
(1) Some of the blood entering the right atrium flows directly to left atrium through the foramen ovale.
(2) Most of the blood from the right ventricle flows through the ductus arteriosus to the aorta.
(3) Very little blood flows through the lungs.
(4) The umbilical artery branches from the aorta.
(5) Blood in the umbilical vein goes to the inferior vena cava.

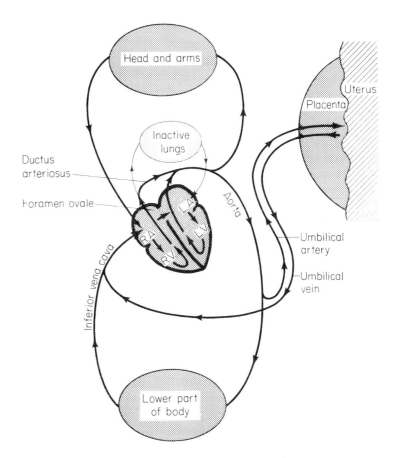

The Change to Air Breathing

As the blood supply from the placenta becomes cut off, there is an increase in carbon dioxide and a decrease in oxygen in the baby's blood, and at the moment of birth the baby is subject to a whole new range of stimuli caused by changes in temperature, touch, sound, light, and freedom of movement. All these factors combine to trigger the respiratory centre in the brain into action. Impulses are sent to the diaphragm and intercostal muscles which then contract; this has the effect of enlarging the thoracic cavity and air rushes in to fill up the space. The respiratory centre then sends out impulses which relax those muscles, so the thorax gets smaller, air is squeezed out—and breathing has begun. The baby often cries with its first breath as air passes out over the vocal cords.

The infant can manage to survive without breathing for the first few minutes of life by using the glycogen stored in its muscles to obtain energy; the glycogen is quickly converted to glucose and energy is released by the conversion of glucose to lactic acid—a process that does not require oxygen. (It is the same as happens during strenuous exercise so that energy can be gained at a time of oxygen shortage).

147

Maintaining the Body Temperature

Full term babies are born with a store of fat which (a) forms an insulating layer beneath the skin and (b) can be used as an easily available source of heat. Babies always have to work harder than adults to keep the body warm as they have a larger surface area in proportion to their weight and, as their heat-regulation centre is poorly developed, it is necessary to protect babies from the cold. It is desirable for new babies to be kept in a relatively stable temperature which is warm but not hot, and the temperature should not fall below 18°C (65°F).

A baby who cannot keep up his body heat will suffer from **hypothermia** (low body heat) and this can be dangerous.

THE NERVOUS SYSTEM

When first born, the baby responds automatically to certain stimuli; these reflex actions are inborn and they control most of the bodily functions.

Sucking and swallowing reflexes enable the baby to feed as soon as he is born.

Rooting reflex, that is, turning the head in search of the nipple when touched gently on the cheek.

Noise reflex the baby startles at the sound of a loud noise—he flings back his head and throws out his arms and legs.

Grasp reflex the baby automatically clenches his fist around any object placed in his palm.

Walking reflex the baby makes automatic walking movements when held upright with the feet touching a firm surface.

Some of these reflexes are necessary for survival (rooting, sucking and swallowing), others seem to be more appropriate to an earlier stage of human evolution: for example, the grasp reflex would have enabled the young of man's tree-living ancestors to cling to the mother or a branch of the tree.

There is a general tendency for the above reflexes to disappear by the age of three months and to be replaced by voluntary actions or conditioned reflexes. At this stage the baby is starting to understand what he wants to do and beginning to be able to do it. From then on he is busily learning many skills. The order in which things are learnt is remarkably consistent but the speed of learning varies from baby to baby.

NEW-BORN BABIES (those up to 4 weeks old):

> do not have a stable sleep pattern.
> eat by sucking.
> have an inborn ability to communicate by sound—they make distinguishable cries to indicate when they are hungry, in pain or lonely.
> are aware of light, although not able to see in the way in which adults can.

are able to hear; they startle at a loud noise and soon learn to distinguish their mother's voice.

are sensitive to smell, they turn their head away from an unpleasant smell.

are born with a degree of muscle control; they can make quite precise head and arm movements.

are born with a degree of co-ordination between the senses. They soon turn their eyes towards a sound source.

have the capacity for rapid learning.

THE BASIC NEEDS OF A CHILD

Whatever sort of home a child comes from, whether rich or poor, Asian, American, African or European, he (or she) will have the same basic needs in order to grow and develop to the full potential.

1. Food The right sort of food in the right proportions. This is discussed in more detail under breast-feeding p. 88, infant and school child p. 102.

2. Warmth Sufficient warmth to keep the body temperature at about 37°C.

3. Protection from both physical damage and from disease.

4. Love A child needs to be loved so that he can learn how to love, that is, how to make long-lasting bonds of affection that are deeply rewarding.

5. Stimulation A child has an inborn zest for learning and if he is given the right sort of stimulation at the various stages of development, his abilities are much more likely to develop to their full potential. For example, he needs the stimulation of being talked to in order to learn words and how to use them.

6. Security A child who is secure in the knowledge that he is wanted by other people acquires self-confidence, and this enables him to overcome the problems and difficulties and challenges that are all part of human life.

7. Personal significance He needs to be treated as though he matters in order to acquire self-esteem.

8. Companions of his own age Playing with other children provides a basis of shared experience which will enable him to make friends—and children need friends.

9. Scope for self-expression Each person needs to develop in his own way—physically, emotionally and intellectually.

10. Discipline Discipline which is firm but kind is essential for emotional development; it teaches self-discipline and allows individuality and personality to flourish.

DIFFERENCES CAUSED BY SEX

By the seventh week of pregnancy male and female differentiation has begun. If the Y chromosome is present the testes begin to form and to secrete hormones which cause the development of the rest of the male genitalia. In the absence of the Y chromosome, the child develops as a female under the control of female hormones.

Individuals show tremendous variation and the following information must be regarded as what generally happens in the majority of cases.

149

DIFFERENCES BETWEEN MALE AND FEMALE PATTERNS OF GROWTH AND DEVELOPMENT

(a) Boys tend to be slightly larger at birth.
(b) Puberty is roughly two and a half years later in males.
(c) Males show later proficiency in a number of intellectual skills.
(d) Bone ossification is completed later, giving boys a longer time to grow in height.
(e) There is wider variation amongst males than amongst females in most aspects of development, for example, there are more idiots and geniuses amongst males than amongst females.

Males are More at Risk

(a) More male foetuses are miscarried (naturally aborted).
(b) Male infants are more susceptible to complications at birth and during the following weeks (prenatal and postnatal complications).
(c) Throughout life men are more subject to some diseases, and to accidents.
(d) Although modern medicines and improved living conditions have lengthened the life span for both and women, the expected life span for men is still shorter than that for women.

Males Have Differing Physical Abilities

Males are physically equipped for greater athletic performance: they develop proportionately larger hearts, lungs and muscles than females.

(a) **Larger muscles** give greater strength.
(b) **Greater stroke volume**; the larger heart allows more blood to be pumped with each heart-beat.
(c) **Greater vital capacity** (p. 38) allows more oxygen to be extracted and carbon dioxide to be excreted.
(d) **More red cells** in the blood give an increased oxygen-carrying capacity.
(e) **Greater calorie intake**

> enables the muscles to produce more energy.

(f) **Less fat and more muscle** give greater speed of movement and less inertia.
(g) **Greater length of arm and smaller carrying angle** gives the ability to throw objects harder and further or hit them with greater force.

Males have a higher metabolic rate than females of similar age and therefore need more food to supply the higher energy requirement.

Causes of Differences between Males and Females

Genetic factors are responsible for:
the sex of the individual,
the functioning of the sex organs,
the differing rates of growth and development,
differing sensory capacities.

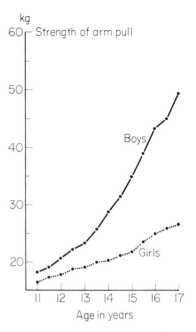

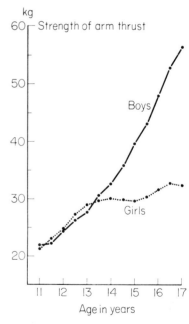

Fig. 15.3 Strength of arm pull and arm thrust from age 11 to 17. Mixed longitudinal data, 65 to 95 boys and 66 to 93 girls in each age group.

Differences in Behaviour

At one time it was taken for granted that genetic factors were responsible for differences in behaviour—they were the underlying cause of the differences in interests, emotions, attitudes and achievements of males and females. Nowadays some people argue that these differences are the result of training—that parents and teachers have one set of expectations for boys and one for girls, and the children respond accordingly (these are environmental factors).

As happens in other aspects of development, the differences in behaviour of boys and girls are most likely to be due to the interaction of genetic and environmental factors, with some cultures preferring to emphasize the male and female differences, and others to minimise them. The traditional behaviour patterns may be described as follows:—

Boys tend to be more
 active and energetic
 aggressive
 assertive
 ambitious
 competitive
 exploratory
 interested in things rather than people
 effective with inanimate objects.

Girls tend to
 be more sensitive to pain
 have a keener sense of smell
 be better at sound discrimination and localisation
 have superior memory in recall and recognition, and be better with words, especially at a younger age
 walk, talk and acquire bowel and bladder control at a younger age
 be more interested in people than things
 be more timid and easily inhibited by novelty and uncertainty
 construe personal relationships and social situations in a more complex manner
 relate events and social situations to themselves rather than in a general context.

GROWTH

When an egg is fertilised by a sperm, a new individual begins to grow and develop. The fertilised egg contains all the instructions to produce a new human being. Shortly after fertilisation, the egg divides first into two, then into four, and so on. Cell division continues throughout life to produce cells for growth, repair and replacement.

During the first three months after fertilisation, the cells are being organised into tissues and organs and very little obvious growth takes place. From then onwards the rate of growth increases and becomes fastest just before and just after birth. At birth the baby weighs, on average, about $3\frac{1}{2}$ kg ($7\frac{1}{2}$ lb) and measures about 54 cm (21 inches), and by six months the baby will have doubled its birthweight.

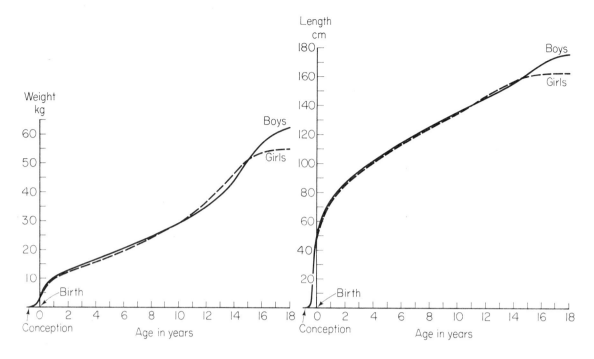

GROWTH FROM BIRTH

If the general rate of growth is plotted on a graph the growth curve shows four distinct phases:

1. A rapid increase during infancy, especially during the first year.
2. Slow, steady growth during childhood.
3. A sharp increase in the rate of growth during puberty.
4. A slow increase during the latter part of adolescence until growth in height ceases and the child becomes an adult.

The growth curve varies for the sexes; the growth spurt at puberty occurs about 2½ years earlier in girls than boys and growth is completed at an earlier age. On average girls are 15 cm (six inches) shorter than boys when they have finished growing.

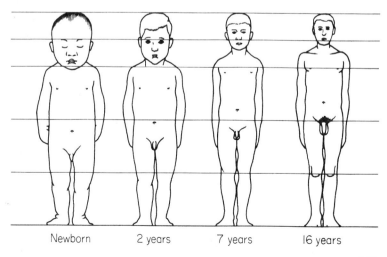

Newborn 2 years 7 years 16 years

Fig. 15.4 (a) Graph to show the relationship between average weight and age for male and female children.

15.4 (b) Graph to show the relationship between average height and age for male and female children.

Note

Growth can be measured by increase in height and increase in weight. Growth curves show distinct sex differences; at birth, on average, boys are slightly heavier and longer than girls and they remain so during childhood. Then, because of earlier puberty, girls temporarily overtake boys at this stage. At adolescence, any individual child will show a sharper increase in the rate of growth than the average child depicted here, and the growth spurt is likely to occur either earlier or later.

Fig. 15.5 (also previous page) Diagrams to show the changing proportions of the body as growth proceeds.

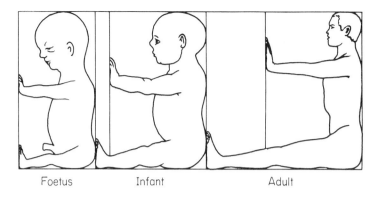

Foetus   Infant   Adult

The rate of growth is not uniform throughout the body. The different tissues grow at different rates and certain parts of the body have their own growth patterns; these are reflected in the changing shape and proportions of the body from embryo to adult (Fig. 15.5).

Brain

Compared with an adult, the child has a relatively large head in proportion to the rest of its body because of the early growth of the brain (not the face). At birth the brain is already a quarter of the adult size and it has all the nerve cells (neurones) it will ever possess; it is thought that all the neurones need to be present at this stage because learning begins immediately after birth.

During the first year the brain increases to three-quarters of the adult size due to:

increase in the size of the neurones

increase in the number of dendrites and with a more elaborate branching pattern

development of the myelin sheath around the axons.

Growth then proceeds more slowly due to continued myelination, and the adult size is reached by the age of 17, although it is almost complete by 7–8.

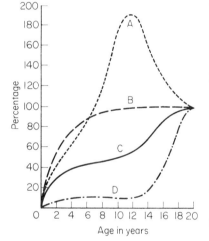

Fig. 15.6 Growth curves of different types of organs. A = Lymphoid type, B = neural type, C = general type, D = reproductive type.

Lymphoid Tissue

Is found in lymph nodes, thymus, tonsils, adenoids. The amount of this tissue increases rapidly until the child is about 12 and then decreases to the adult level. It has this pattern of growth because childhood is the time when the body is rapidly acquiring immunity to a large number of pathogens and needs to make a large number of antibodies.

Reproductive Organs

These show little change until the beginning of puberty when rapid growth takes place and continues throughout adolescence.

Teeth

The growth of teeth begins in the womb and continues throughout childhood and adolescence until the wisdom teeth come through during the teens or later. At birth the milk teeth are present in the jaw and the permanent teeth are already starting to develop. Teething is under genetic control and cannot be hurried up or delayed. The first tooth, often a lower incisor, usually appears at about six months, but it may be much later, with the other incisors following, and the premolars erupting before the canines. See also p. 42.

THE LONG CHILDHOOD

During this long period of slow, steady growth the child remains relatively small and weak and under the care and control of the parents. The child has a natural zest for learning and the long childhood enables him to be taught and to learn:

to communicate by speech
to use his hands and develop complicated neuro-muscular skills
to co-operate with other people
to absorb the attitudes of the community in which he is growing up.

FACTORS WHICH INFLUENCE GROWTH

1. **Genetic Factors**

The genes which the child inherits from his parents will determine the maximum height to which he can grow. Adverse environmental factors may prevent him from growing to his full potential height but favourable environmental factors will not help him to increase it.

2. **Environmental Factors**
   (a) **Nutrition** Severe malnutrition has a considerable effect on growth, affecting weight more than height, and may be due to poverty, ignorance, food fads or disease. It may stunt intellectual growth as well as physical growth.
   (b) **Illness** Severe illness may retard growth and the lower the age of the child the greater the risk of its having a permanent effect. If a child's normal rate of growth is only temporarily slowed down by illness he will adapt afterwards with a period of faster growth until he has caught up.
   (c) **Season** The rate of growth in height is affected by the seasons, being faster in spring and slower in autumn. Children gain more weight in winter than summer since they are less active and eat more in cold weather.
   (d) **Stress** Due to overcrowding, unhappiness etc. can retard growth in height and affect weight by causing thinness or fatness.

DEVELOPMENT

Development means an increase in complexity and is more difficult to assess than growth. Development is a continuous but not a smooth process; it tends to go forwards in spurts, sometimes in one direction, sometimes in another.
Different fields of development include
Neuro-muscular control
(a) posture and large movements
(b) hand–eye co-ordination
(c) bladder and bowel control
Language
Vision

154

NEURO-MUSCULAR CONTROL (the co-ordination of nerves and muscles)

The nervous system has to learn how to control the muscles responsible for body movements. This takes place alongside growth, and better co-ordination and more complex skills can only be acquired when the body has matured enough to have the physical ability to carry them out.

Movements are at first clumsy,

they gradually become finer,

the child becomes able to perform movements more rapidly and with increasing accuracy,

he can then tackle tasks of increasing complexity demanding higher levels of skill.

When a movement is carried out, a nerve impulse will have travelled along a particular pathway through the nervous system. The first time that the movement was made, the impulse might have had difficulty in finding the correct pathway along which to travel, but doing it once makes it easier to do it a second time, and practice will make it easier still. This is the way new reflexes are acquired—**conditioned reflexes**—which are acquired by learning. The development of conditioned reflexes is involved in learning to walk, talk, swim, write, etc.

The physical development of a child in the first year of life is under genetic control and seems to take place largely independently of environmental influence.

Posture and Large Movements

There is progressive development of muscular control from head to toe. Control of movement starts with the head and continues down the trunk, arms and legs. The ability to lift the head is followed by the ability to hold the head erect, sit supported, sit unsupported, crawl, stand, walk, run, jump, hop and so on.

Hand-eye Co-ordination

There is progressive development from stretching out for objects, to grasping objects with the whole hand, use a finger and thumb hold, use a spoon, start to undress, and then to dress, to use manipulative tools such as scissors, and so on.

BLADDER AND BOWEL CONTROL

Young babies have no voluntary control over the passing of urine and faeces. If a baby of a few months old performs on the pot this will be due to:

(a) reflex action when the cold rim of the pot touches his bottom; or

(b) he has regular bowel movements at a particular time of the day.

Babies commonly empty the bowel and bladder immediately after a meal. Most babies who use the pot in the early months stop doing so around 9–12 months and will not be able to do so again until they have learnt to control the sphincter muscle at the outlet to the bladder and the muscles of the anus (anal sphincter) at the outlet to the bowel.

155

Normal Development of Bladder Control (control of micturition)

1. The first stage is awareness of having passed urine, at above 15–18 months.

2. The child then begins to tell his mother just before he does so.

3. When he has acquired some control he is able to tell his mother when he wants to pass urine and to wait until he has been put on the pot.

4. Most children take responsibility for their own toilet needs during the day by the time they are $2\frac{1}{2}$ years; often they are not dry at night for some while, possibly this is related to the size of the bladder. Bowel control (control of defaecation) is likely to be established a little earlier than bladder control and is acquired in a similar manner.

NAPPY RASH

Nappy rash and sore bottoms are extremely common during the nappy-wearing stage and are caused by the skin being in contact with urine and faeces. Breast-fed babies suffer less than bottle-fed babies because the faeces are more acid and discourage the germs responsible for soreness.

Prevention

1. If the skin begins to look sore, clean it gently with oil or baby lotion.

2. Ointments such as petroleum jelly or zinc and castor oil form a barrier between the skin and a wet nappy.

3. Exposure to the air helps the skin to recover.

4. A 'one-way' nappy may help.

5. Keep the nappies soft to prevent chafing by washing in pure soap and rinsing well to get rid of all the soap. Detergents and enzyme washing powders may cause irritation.

LANGUAGE DEVELOPMENT

Babies have an inborn ability to make noises, the desire to communicate by making sounds, and the brain capacity to learn how to turn these sounds into words and to use words to make sentences.

How a Baby Communicates

1. Cries to indicate hunger, discomfort or loneliness and makes little guttural noises to indicate contentment (new-born).

2. Eye to eye contact (one month).

3. Smiles to show pleasure (5–6 weeks).

4. Vocalises freely and likes to hold 'conversations' (3 months).

5. Laughs and chuckles when happy and screams when annoyed (6 months).

6. Indicates what he wants by pointing (12 months).

7. Begins to learn how to turn noises into speech (12 months onwards).

Babies learn to talk by:

   (a) hearing other people talk

   (b) practising making noises

   (c) imitating sounds made by other people

(d) practising these sounds until they become automatic

(e) learning what the sounds mean; babies often understand the meaning of words long before they can say them.

Pattern of Language Development

Although language development follows a pattern there is wide variety in the speed at which children learn to talk. Some will be in advance of, and some behind, the average ages mentioned here.

The new-born baby is able to cry and make little guttural noises when content. He soon learns how to make other sounds and it gives him pleasure to use these to hold 'conversations' with people (3 months).

He spends much time practising these sounds using single syllables e.g. ka, goo, der, and then adds double syllables to his repertoire e.g. adah (6 months). He learns to copy sounds made by other people and comes to understand that these sounds have meaning. By one year old he understands the meaning of several words and may be able to say one or two. The baby continues to learn more words and to practise by talking to himself, at times continuously. At two years old he may be able to put 2–3 words together to form a simple sentence.

He next learns to use pronouns, I, me, you and by three can carry on simple conversations and ask many questions. Speech is usually completely intelligible by the age of four, and by five is both fluent and grammatical.

Delay in Talking May be Due to:

1. **Genes**   There may be a family pattern of being late to talk.

2. **Concentrating first on other aspects of development** such as walking.

3. **Lack of stimulus**   If no-one bothers to talk to the child he will not be able to imitate or to understand words. This reduces the speed and quality of a child's development of language.

4. **Lack of motivation**   If no-one shares the baby's pleasure in the sounds he is producing or encourages him to produce different sounds.

5. **Deafness**   It is usual for a deaf baby to make the usual sounds of gurgling and babbling, but speech development does not progress any further as the deaf infant does not hear noises to imitate. It is extremely important to recognise deafness in children at an early age, preferably under a year, so that specialist teaching can be given. A deaf child who does not hear sounds until the age of three is slow in learning new sounds and, if he has not heard them by the age of seven, it is almost impossible to teach him the sounds.

There is a danger that if deafness in a child is not recognised, the inability to learn to talk will be wrongly attributed to being mentally retarded.

VISION

A new-born baby does have some vision and there is some evidence that he is born with the ability to understand a three-dimensional world, but he cannot see in a clear way.

157

At birth he is aware of bright light and will turn his head and eyes in that direction.

Quite soon he is aware of fairly large objects 18–25 cm (7–10 inches) from his eyes as objects at this distance form the sharpest image (young infants are extremely short-sighted), and he gazes at his mother's face as she feeds him with an increasingly alert expression.

Within a few weeks he can detect differences between faces and is able to recognize his mother.

At three months he is visually very alert and particularly interested in human faces. Now that he can see further he actively watches people and follows them around with his eyes. When objects move close to his face he is now able to blink defensively, and his eyes converge.

At six months his eyes have learnt to move in unison (binocular vision) and to squint is now abnormal as he can easily focus small objects within 15–30 cm (6–12 inches), and he can also watch moving objects at 300 cm (10 feet).

At a year he can recognise people 600 cm (20 feet) or more away.

At 3 years he can match up two or three primary colours, usually red and yellow, but may confuse blue and green.

At four years he can match and name four primary colours correctly.

At five years he can match 10 or 12 colours.

FEET
Functions
1. to balance and support the weight of the body,
2. to act as levers to propel the body forwards in walking, running and jumping.

Growth

At birth, the bones of the feet are mainly of cartilage, and this is gradually replaced by bone during childhood and adolescence. Cartilage is pliable and young feet can easily become deformed if they are crushed into shoes which are too tight, although the child may not be aware of much pain.

It is normal for babies to appear to have flat feet, because the arch is filled with a pad of fatty tissue which disappears when the child learns to walk. Walking strengthens the muscles which hold the bones in position and move the foot, and walking barefoot allows the muscles and bones to develop in the natural way to produce strong healthy feet.

Shoes—are worn to protect the feet against cold and other damage and they need to be well-fitting to allow the bones and muscles to develop properly. Points to look for when buying children's shoes are:

1. Flexibility Flexible shoes that bend easily with the foot encourage a natural springy step. If the shoes are stiff, the child's muscles will not be strong enough to bend them, and consequently the muscles cannot be used in the proper way for walking; this leads to bad walking habits and posture.

2. The correct size Shoes need to be wide enough and long enough so that the toes are not cramped and the big toe can lie in its natural position.

If shoes are too big, the child will slouch along and will not be able to exercise and develop those muscles that give spring to the step.

3. Room for growth This is important, especially for younger children, as the feet grow fast as soon as the child begins to walk; the recommended growing space is about 18 mm ($\frac{3}{4}$ inch).

Socks Tight socks can also do harm. They should be big enough to give a loose easy fit at the toe.

Fashion shoes Feet are not fully formed until about the age of twenty, so adolescents who wear shoes which put unnatural strain on bones and muscles run the risk of causing serious damage to their feet. In particular, shoes with pointed toes encourage the big toe to lie across the next toe, leading to bunions and osteoarthritis of the big toe joint, also hammer toes, clawed toes and painful corns.

Damage to feet is long-lasting and worsens with age; many people, especially the elderly, suffer because:
the feet are painful and it is very difficult to find comfortable shoes.
painful feet affect posture, possibly putting strain on the spine.
mobility and independence are reduced if walking is difficult.

Bunions Tight shoes cause the big toe to be pushed towards the other toes so that the head of the metatarsal protrudes, and the joint between the big toe and the metatarsal becomes enlarged and painful.

Corns Continuous pressure on a small area of the skin makes it produce a mass of horny, dead tissue and this causes pain.

Blisters Continues rubbing on one area of skin makes fluid collect underneath the epidermis and irritates the nerve endings in the dermis, causing pain.

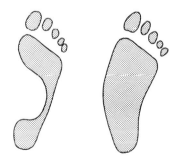

Fig. 15.7 Left, normal foot print; right, flat foot.

Flat feet and foot strain are caused by standing still too long. Standing still for a long time makes the soles of the feet ache because the muscles become tired and so allow too much strain to fall on the ligaments. If this happens regularly, the ligaments become stretched and cannot recover; the result is 'fallen arches' and flat feet. Treatment for flat feet which ache is to strengthen the foot muscles by exercise, and, if necessary, to reduce the weight which they have to support, and to wear shoes which give support to the arches.

Flat feet in children only need special attention if the feet are painful or stiff or if the deformity is the result of some other condition. Almost all children under 18 months of age appear to be flat footed, whereas amongst ten year olds, it is an uncommon condition. This means that the 'flat foot' of the baby usually corrects itself as the child grows.

159

ADOLESCENCE

Adolescence is the time of transition from childhood to adulthood. It begins with the growth spurt which precedes puberty and during this time boys and girls choose friends mainly of their own sex, but as adolescence proceeds, they show more interest in the opposite sex and want to meet and mix with them.

Hormones   During adolescence the body is adjusting to an altered system of hormones. Hormones govern the **physical changes,** are involved in the **emotional readjustments,** and play some part in the **altered patterns of behaviour** characteristic of adolescence.

The Physical Changes of Adolescence

(a) Acceleration of the rate of growth in height.
(b) Puberty takes place during the early part of adolescence and is the time when the sex organs mature and become capable of producing sperm or ova.
(c) The secondary sexual characteristics develop (p. 89).
(d) There is an increase in output of sebum from the sebaceous glands in the skin and this often leads to acne; it happens to many people and can be considered a normal part of adolescence. Nowadays it is possible to treat this condition and it is not necessary to endure years of embarrassment.

Emotional Readjustments

Adolescents have to:
(a) cope with their changing sexual feelings,
(b) make the transition from being dependent (on parents) to being independent (of parents).
(c) adjust to changes in behaviour, attitudes and feelings shown them by other people because they are no longer children.

Behaviour Characteristics of Adolescents

Not all adolescents exhibit all of the following, but most show at least some to a greater or lesser degree:
fatigue and lassitude
rapid emotional ups and downs
sensitiveness about appearance
lack of confidence
awkwardness and clumsiness
desire to conform with others of their own age group (their peer group) in appearance and behaviour
refusal to be bound by the opinions of adults
desire for independence, leading to conflict with parents over such matters as staying out late or leaving home.

The Parents' Point of View

The rapid emotional ups and downs are rather trying to live with.
It is hard to know when to relax discipline and accept that their child has his own point of view and his own life to lead.
They are aware of the dangers that befall the inexperienced.

The problems of adolescence are likely to be emotional rather than

160

physical, and generally, they are problems which have their roots in childhood but have remained under cover. For example, the inability of parents and their adolescent children to talk over matters of importance is much less likely to arise in families where the parents had got into the habit of talking to their children during earlier years and to **listening** to what they had to say.

PARENTHOOD

With the knowledge of contraceptive techniques readily available, it is nowadays possible to decide whether or not to become parents. **It is very important** that parents should be reasonably happy together and fairly self-confident because:

children are hard work. When young they are by nature neither clean nor tidy nor responsible, and for the first few years of life they are completely dependent on their parents; after that they only slowly become independent and capable of looking after themselves.

they grow from 'dear little' babies into children and adolescents who are much more difficult to manage.

they can bring great problems, especially if they are unloved, unwanted or insecure.

they require sacrifice from the parents of both time and money.

they bring change to the parents' lives, restricting their freedom but at the same time providing a new dimension.

they can bring great joy to the family if the parents are prepared to spend time willingly on bringing them up.

GUIDE LINES FOR PARENTS

1. **Love your baby**   } to give him or her the security
2. **Cuddle your baby** } of being loved and wanted.
3. **Talk to your baby** so that he can learn how to speak.
4. **Listen to your baby,** as well as talk to him, and you will learn how to communicate with each other. Many of the problems of adolescence are due to lack of communication between parents and child.
5. **Play with your baby,** he will enjoy it, and it will enable a far closer relationship to develop between you.
6. **Be firm with your baby** when he is old enough to understand what is wanted of him. There is no point in asking too much of a baby or young child as it will only make you both unhappy; he only very gradually understands that his mother is pleased by some of the things he does and not by others.
7. **Reward is more effective than punishment** in the training of children; if a child only gets attention when he is naughty, he is unlikely to see any point in being good.
8. **Keep your baby clean,** but not too clean; children need the opportunity to come into contact with pathogens (germs) because childhood is the time when immunity is easily developed against many of the diseases which are dangerous to adults.
9. **Do not over-protect.** Children need to learn gradually to be responsible for their own lives.
10. **Do not spend so much time and energy** on your children that you do not have any left over for yourself and your own interests.

Remember that children grow up and gradually need to become independent.

No wonder child-rearing is a long, hard job. This is why it is often considered undesirable for young people still in their teens to become parents; they themselves have not finished growing up and may not be sufficiently mature to cope with the demands of parenthood.

Preparation for Being a Parent

This begins in childhood by playing with dolls or playing 'mothers and fathers'. How the parents themselves were brought up plays an important part, and they also learn by helping to bring up younger brothers and sisters or cousins or neighbours' children.

In these days of smaller families it is quite possible that parents will have very little knowledge of how to bring up children and, if they live away from their own families, then the grandparents will not be at hand to give help and advice.

Every family with a new baby is visited by the Health Visitor and she will give advice and support where necessary.

The importance of the family

To feed, clothe and protect the child.

To provide a setting where he can develop as a unique person.

To provide a secure base from which he can venture in order to learn more about the outside world and ultimately to establish a place there for himself.

To teach the child the values of the cultural group to which he belongs.

MIDDLE AGE

People age at different rates and how old a person feels is very much a state of mind, but generally, middle age is most often thought of as the years between forty and sixty. For many it is a time of:

learning to be the parents of adolescents,

becoming grandparents,

being nursemaids to elderly sick parents,

for some people, getting to the top of the ladder, for others the realisation that they are never going to get any further,

preparation for retirement.

Middle age is the 'prime of life' for many people because:

they have learnt to come to terms with themselves and know their own strengths and weaknesses,

they can be of great service to society through their experience, knowledge and maturity,

it can be the beginning—perhaps of a new career, a return to studying, a chance to take up hobbies and interests again, or to travel.

FACTORS AFFECTING HEALTH

Health in middle age is due to a combination of factors.

Tissues and organs undergo gradual change throughout life and

162

during middle age people become aware of some of these changes such as as in eyesight and muscle power,

Parentage; some health problems have a tendency to run in families, for example, varicose veins and sugar diabetes.

Childhood; for example, children who suffer acute (short-term) bronchitis are more likely to develop chronic (long-term) bronchitis in later life, and ill-fitting shoes in childhood are the cause of many of the foot troubles of older people.

Decisions taken 10, 20 or 30 years earlier about, for example, smoking, diet and exercise.

Physical Changes in Middle Age

Sight   The eye gradually alters with age and one of the changes is to the lens. From childhood, it gradually becomes less flexible and hence more resistant to alteration in shape in order to bring near objects into focus; the change to the lens becomes obvious when a paper has to be held at arm's length in order to read it, and this often happens in middle age. With increasing age, the lens becomes more flattened on both sides until it becomes a fixed-focus lens— focused on distant objects. (Change to the lens is why short-sighted people find their sight improving with age).

Hearing   Throughout adult life, the ear gradually becomes less sensitive, particularly the part that detects high notes. In addition, hearing can be damaged by continuous, excessive noise, e.g. from discos, radios, factories, noisy roads.

Skin   As a person grows older, the skin becomes less elastic, drier, and loses the thin, underlying layer of fat, and wrinkles develop. The process is accelerated by exposing the skin to sun and wind. Skin cream is useful if used

   to protect the skin against the weather,

   as a barrier against chemicals such as detergents,

   to soften and prevent water loss (drying-out) from skins.

Muscles   Gradual loss of muscle power is noticed when greater effort is required to perform the same task; very energetic sports like football are replaced by gentler games like golf and bowls.

Obesity   There is a tendency to put on weight in middle age because, whilst the appetite remains constant, less exercise is taken to use up all the calories.

Varicose veins   A tendency to develop varicose veins often runs in families and they are encouraged by standing for long periods, lack of exercise, overweight and pregnancy. Apart from looking unsightly, they cause the legs to ache and ankles to swell.

Menopause   When the child-bearing years are over, the female hormones (oestrogen and progesterone) are no longer made in such large quantities; eggs are no longer produced and the periods finish. During the menopause the female hormones are readjusted to the girlhood pattern; it is like puberty in reverse, and takes about the same amount of time for completion, that is, five years.

In some women the change to the hormone pattern happens easily and there are no side effects; in others the hormones temporarily get out of balance and cause hot flushes, tiredness, irritability, changes of mood, and so on. Many of these symptoms can have emotional as well as physical causes, and be due perhaps to loneliness, unhappiness, depression, or the difficulty in readjusting to the children becoming independent. Medical help can alleviate these symptoms and there are Menopause Clinics which specialise in this field. Hormone Replacement Therapy (HRT) is one form of treatment—it is the replacement of natural oestrogen no longer being made in the body by oestrogen tablets.

Although men do not have a menopause, the male hormone testosterone gradually declines, in some men more rapidly than others; the number of sperms produced gets fewer, but elderly men may still be able to become fathers.

Cancer of the breast and cervix Screening procedures exist to detect these diseases and both can be cured if caught in the early stages. Women, especially those in middle age, should examine their breasts once a month between periods, and report any change to the doctor. A cervical smear should be taken every few years—a few cells are painlessly removed from the cervix and examined under a microscope.

Softening of the bones After the menopause, women start to lose calcium (the hardening material) from their bones; it is a long-term process and takes place more slowly in some women than others.

AGEING

Ageing is a natural process and results in gradual changes in the various tissues and organs which prevent them from functioning as efficiently as they used to.

NORMAL PROCESSES OF AGEING

Collagen fibres (white fibres) weaken; they form the scaffolding between the tissues of the body and are the principal component of connective tissue.

Elastic fibres (yellow fibres) become hard and no longer allow such parts as the skin to stretch and then return to size.

Skin becomes more wrinkled and drier and lacks a fat layer under the surface.

Baldness The hair usually grows thinner and greyer, and may be lost altogether from the top of the head, especially in men, although it does happen to a few women.

Height It is usual to lose height, mainly because the inter-vertebral discs get thinner.

Posture The bent posture of many old people may be due to habit and not to disease (have they got painful feet?).

Movements become slower due to reduction in the speed and power of muscular contractions, and it also becomes impossible to sustain hard muscular work for so long.

Joints stiffen if
(a) the muscles stiffen,
(b) **Osteoarthritis** (osteoarthrosis) affects the joints; the articular cartilage degenerates and may be replaced by rough deposits of bone tissue; the joint creaks and movement is painful. The most susceptible joints are the weight-bearing joints of legs and spine, and movement at these joints is made worse by overweight and faulty posture.

Osteoporosis weakens the bones and they break more easily; this condition is more common in women than men. Bones become

164

weakened due to loss of strength of the collagen fibres and to a shortage of calcium compounds.

Hardening of the arteries—arteriosclerosis (atherosclerosis) The arteries lose their elasticity and their walls become thickened by the deposition of fatty material on the inside. These conditions have the effect of reducing the amount of blood that can flow through them and this may impair the efficiency of the organs they serve. If the artery walls become damaged and a blood clot forms, the clot will cause a 'heart attack' if it blocks the blood supply to the heart muscle, or a 'stroke' if it blocks the blood supply to the brain.

Tendency to slower mental function   Although from an early age the number of brain cells steadily decreases, this does not necessarily have a marked effect on the efficiency of the brain as there are still many millions of brain cells left. There is however some evidence that old nerve fibres conduct impulses more slowly and this will result in slower reactions. If the arteries to the brain harden and thicken to such an extent that the blood supply to the brain is reduced, the brain cells will suffer from a shortage of oxygen and mental efficiency will be impaired.

Tendency to shortage of breath (breathlessness). The vital capacity (p. 38) of the lungs steadily falls as the muscles controlling the size of the thorax (p. 38) lose their strength and as the lung tissue becomes less elastic, and slight exertion may result in shortage of breath.

Tendency to loss of vision—presbyopia (p. 72).

Tendency to deafness   Throughout adult life the ears become less sensitive to high-pitched notes. Low sounds and loud sounds can be heard but other people's speech becomes more difficult to understand because the high sounds of the consonants (1, t, d, etc) cannot be distinguished. Shouting at a person with this form of deafness does not help but speaking slowly and clearly does; he will be able to hear more easily through a telephone because this transmits low sounds. It should be remembered that much deafness is caused by wax in the ears.

Tendency to loss of smell   When this happens people will not be able to detect such things as escaping gas or smelly conditions in the house. They will also lose some of their sense of taste as this partly depends on smell.

Difficulty in keeping warm   Slower movements mean that less heat is generated by the muscles, and the temperature-regulating mechanism may not work as efficiently. Old people (like babies) are subject to **Hypothermia.**

The rate   at which people age depends on many factors—place, climate, heredity, overwork, overweight, malnutrition and so on. Some very old people remain energetic, mentally active and enjoy life, whilst other much younger people are 'old for their years'.

TO HELP DELAY THE EFFECTS OF AGEING

1. Keep physically active for 'What you don't use—you lose'
   (a) muscles deteriorate if they are not used,
   (b) movement aids circulation and good circulation keeps the cells well supplied with food and oxygen and removes waste products.

2. Keep mentally active—a brain that is not used deteriorates.

3. A well-balanced diet is required in order to keep healthy (p. 102).

4. Avoid becoming overweight; obesity leads to or aggravates many disorders such as high blood pressure, heart disease, diabetes, osteoarthritis, and it may be the cause of difficulty in walking far or climbing stairs.

5. Take care of the feet Much pain and disability can be avoided by preventing or treating corns, bunions, callouses, hammer toes and long toe-nails (p. 159).

6. Be interested in other people rather than yourself and your own health—it helps prevent loneliness.

GROWING OLDER BRINGS BENEFITS

Life becomes less emotional and more tranquil.

There is time to pursue interests and hobbies.

There has been time to gain in wisdom, experience and understanding.

A lifetime's experience is why the young often turn to older people for advice.

It is usual for skills that do not require hard muscular work to remain with older people if they continue to use them, for example, piano-playing and needlework, and they can give much pleasure.

Although learning is more difficult, it is still possible to acquire new skills that there may not have been time or opportunity to learn before.

Chapter Sixteen

ENVIRONMENTAL HEALTH

Environmental health has been defined by the World Health Organisation as the control of all those factors in man's physical environment which exercise, or may exercise, a deleterious effect on his physical development, health and survival.

The importance of hygiene and sanitation have been recognised since earliest times but, in this country, laws were not made on such matters until the middle of the 19th century. In the latter half of the 18th century the industrial revolution had caused the migration of people from the countryside to the towns, where they lived and worked in over-crowded and unhygienic conditions. Gradually, it became apparent that some form of control over their living and working conditions was necessary. Action was taken following a cholera epidemic in 1834 and an outbreak of typhus in London in 1837, which together with a bad harvest, rising prices and unemployment caused the government to set up an enquiry. This enquiry was conducted by three doctors who exposed the dreadful living conditions and the amount of preventable disease in London. From then onwards parliament gradually passed Acts to control working hours, water supplies, sanitation, housing, factories, etc.

As scientific knowledge of disease increased these measures became more effective.

Some important dates:

1842. Mines Act prohibited the employment of women and of boys under 10 years underground.

1844. Factories Regulation Act limited the working hours of women to 12 a day.

1847. First Medical Officer of Health (M.O.H.) appointed—for Liverpool.

1848. First M.O.H. appointed for London.

1859. First District Nurse appointed—for Liverpool.

1862. First Health Visitor appointed by Manchester and Salford Ladies Health Society.

1866. Sanitary Act made it a duty of the Local Authorities to control all nuisances, i.e. that which causes inconvenience or hurt to the public.

1872. Public Health Act made the appointment of Medical Officers of Health compulsory.

1875. Public Health Act consolidated and extended previous legislation.

1885. Housing of the Working Classes Act.

1902. Midwives Act.

1907. School Medical Inspections introduced.

1936. Public Health Act. The second large consolidation Act.

1946. National Health Service Act.

1956. Clean Air Act.
1963. Offices, Shops and Railway Premises Act.
1969. Housing Act.
1970. Local Authority Social Services Act.
1973. National Health Service Reorganisation Act.
1974. Health and Safety at Work, Etc. Act.

ENVIRONMENTAL HEALTH OFFICER

Environmental Health Officers (E.H.O.'s) are concerned with the maintenance of environmental health. Their duties include:

1. The inspection for cleanliness of:
    (a) Shops, offices, factories, cinemas, caravan sites, public swimming pools etc.
    (b) The water supply.
    (c) Refuse disposal.
    (d) Sewage disposal.
2. The inspection and supervision of food (p. 109).
3. The inspection of houses to see if they are fit to live in, verminous, or overcrowded.
4. The enforcement of the Clean Air Act which makes it an offence in certain areas for smoke to pollute the atmosphere.
5. Rodent control.
6. Noise abatement.

PUBLIC ANALYST

The public analyst is a chemist who analyses the samples of food and drink sent to him by the E.H.O. He can detect chemicals which are injurious to health and he checks that foods are correctly labelled.

PUBLIC HEALTH LABORATORY

Samples of milk, ice cream and other foods are sent to the Laboratory to be tested for bacteriological content and keeping quality. Samples from the water supply are tested for bacterial contamination. Faecal samples in suspected cases of food poisoning are sent to the Laboratory for investigation.

HEALTH VISITOR

A health visitor visits people in their homes to advise on matters of health. She must be a State Registered Nurse who has midwifery qualifications and the Health Visitors Certificate.

She is then qualified to advise on such matters as:

1. The care of young children and sick, aged or handicapped people.
2. Health education and the prevention of social diseases such as T.B. and V.D.

Health Visitors co-operate with doctors, district nurses, midwives, E.H.O.'s and social workers.

NATIONAL HEALTH SERVICE

The National Health Act of 1946, established the National Health Service. This was based on the principle that people should pay for the Health Service through taxes and contributions when they are well, so that they can receive comprehensive medical benefits when they are ill. In this way the financial burden of sickness is borne by the whole community.

The National Health Service is under the control of parliament and administered by a government department at present called the Department of Health and Social Security. It is financed partly by compulsory National Health Service Contributions paid by people who are working, but mostly by taxes, e.g. income tax, taxes on petrol, cigarettes, beer, etc.

The National Health Service is concerned with:
Hospitals,
General Practitioners (family doctors),
Dentists,
Opticians,
Pharmacists (Chemists).

The Department of Health and Social Security together with Local Authorities administer:
Health Centres,
Care of mothers and young children,
Midwifery,
Health Visitors,
Home nursing,
Vaccination and immunisation,
Ambulance service,
Care and after-care.
Etc.

The National Health Service Reorganisation Act, 1973, came into force on April 1, 1974, and it has necessitated the complete re-organisation of the National Health Service.

HOUSING

Before the industrial revolution most workers lived in cottages near the farms on which they worked. These cottages were well-built as a protection against the weather and each had a fireplace and a privy. However, with the development of industry, factories were built, and nearby, large areas of low-standard housing for the workers. These small, dark, overcrowded houses were crammed together and although each had a fireplace there were communal wash-houses, wells and privies. Thus the slums were formed, and even today slum conditions still exist in some of our large cities.

During the latter part of the 19th century, it was recognised that the poor health of the community was due to the appalling living conditions of much of the population.

The Housing Act of 1890 was the first real attempt to improve living conditions. The Town Planning Act of 1909 forbade the building of back-to-back houses. The Housing Act of 1919 provided government subsidies for the building of Council Houses which resulted in millions of such homes being built, the aim being to provide good quality houses to rent to low-income families.

The Housing Act of 1957 established the principle that every family had the right to a healthy home. This is considered to be one which

(a) is in a good state of repair.

(b) is free from damp.

(c) is adequately ventilated.

(d) has enough windows to allow sufficient daylight to enter.

(e) has a piped water supply.

(f) has an internal sanitary convenience.

(g) has facilities for the storage, preparation and cooking of food.

One of the duties of the E.H.O. is to inspect housing to see if it is fit to live in. If it does not meet the required standards, the Local Authority must decide whether it should be repaired or demolished. If it is capable of repair or improvement the owner may be entitled to a grant to help with this improvement.

LIGHTING

Rooms need to be adequately lit to prevent eye strain, and stairs and corridors need enough light to prevent accidents.

Living rooms should have windows large enough to admit sufficient daylight to see that the room is clean. A window with both a pleasant view and a sunny aspect has a beneficial effect on mental outlook.

The amount of light required depends on occupation. Sitting rooms need a gentle light, a stronger light is needed in the kitchen and workshop, and an operating theatre needs a very bright light.

VENTILATION

Air out of doors is in a continuous state of motion and therefore the composition remains more or less constant. However, the composition of air in a closed room containing people soon alters because:

1. Body heat and moisture from respiration and perspiration increases the temperature and humidity.

2. The oxygen content falls, and the carbon dioxide content increases.

3. The germ content increases if germs are being breathed out by any infected persons. Most germs can survive for some time in air.

4. If cigarettes are being smoked they quickly pollute the air.

It is the increase in humidity and temperature in an unventilated room that causes discomfort and lassitude rather than the alteration in the amounts of oxygen and carbon dioxide. Also the risk of droplet infection increases.

Ventilation of a room occurs through doors, windows, fireplaces, air bricks, electric extractor fans and air conditioning. Swing windows are useful in that they encourage the circulation of air—the warm, stale air leaves the room through the top of the window and it is replaced by cold fresh air flowing into the room through the bottom of the window.

HEATING

Buildings can be heated in a variety of ways.

1. Coal Fires

Disadvantages

Coal is expensive.

They are inefficient, as 50% of the heat goes up the chimney.

The smoke pollutes the atmosphere.

The ash causes dust in the room.

A coal store is needed.

Advantages

They are cosy and cheerful.

They assist ventilation through the fireplace.

N.B. Smokeless fuels have to be used in some areas which have become 'smokeless zones' to prevent pollution of the atmosphere.

2. Electric Fires

Disadvantages

They have no automatic ventilation.

They are subject to power failures.

Advantages

They are clean.

They are easily regulated, as they can be switched on and off as required.

The room warms up quickly.

3. Gas Fires

Disadvantages

None, but care must be taken when lighting gas fires, and any leakages should be reported immediately to the Gas Board.

Advantages

They are clean.

They are easily regulated.

They assist ventilation, as the room must have an outlet for the waste products of combustion.

4. Paraffin Heaters

Disadvantages

As these heaters are portable, there is danger of fire if they are moved when alight and the paraffin is spilt.

They should be used only in ventilated rooms.

If not kept clean, they produce a distinctive smell.

Advantages

They are a cheap source of heat.

They are portable.

5. Central Heating

Disadvantages

It is expensive to install.

There is no automatic ventilation.

Advantages

Large quantities of heat are produced relatively cheaply.

This is the only satisfactory way of heating large buildings.

An even temperature is maintained which is particularly good for old people, babies, and the sick.

The bedrooms are heated. Cold bedrooms are thought to aggravate bronchitis and other lung complaints.

HEAT INSULATION

Less heat is lost from a building which is insulated by:
1. Double glazing.
2. Cavity walls.
3. Lagging the roof, i.e. putting a layer of insulating material between the joists in the loft.

TOWN PLANNING AND THE IMPORTANCE OF OPEN SPACES

From 1909 onwards the government began to pass laws to control the extent and type of building and to plan the environment.

The Town and Country Planning Act, 1947, required Local Authorities to produce development plans and to allocate land for different uses such as roads, housing, schools, industry and open spaces.

As most of the population live in towns, it is necessary to try to keep a reasonable proportion of built-up areas to open spaces, such as parks and playgrounds.
1. Children need space in which to play.
2. Playing fields provide both children and adults with the opportunity to take exercise by playing games.
3. Exercise can also be taken by walking in the park. This is more beneficial than walking along streets, as the air is less polluted by traffic fumes and the scenery more restful.
4. Besides being pleasant to look at, plants are useful as during photosynthesis they absorb carbon dioxide, which is a waste product of respiration, cigarette smoke, the motor car, coal fires, oil fires and gas fires. The oxygen liberated during photosynthesis restores the oxygen content of air.
5. Open spaces allow sunlight and exposure to kill germs.
6. People exposed to sunlight can make Vitamin D in their skin.

OVERCROWDING IN RELATION TO HEALTH

Overcrowded living conditions are detrimental to health because:
1. It is difficult to keep the premises clean.
2. Inadequate washing facilities means that it is difficult for the people to keep themselves and their clothes clean.
3. Insufficient lavatories are a cause of dirtiness and infection.
4. Germs and other parasites such as lice can spread from one person to another more easily.
5. Ventilation may be more difficult.
6. Overcrowded sleeping arrangements may result in lack of sleep.
7. Too many people confined in a small space may build up tensions and stress which result in mental ill-health.
8. Noise may become a nuisance and cause ill-health in some people.
9. People who live in overcrowded conditions tend to be those who do not know of, or cannot afford to observe, the principles of health (p. 130).

172

FACTORIES

The growth of industry in the latter part of the 18th century caused large numbers of people to be employed in factories. Concern began to be shown about the conditions under which they worked and the 19th century saw the beginning of legislation to improve these conditions.

1802. Pauper apprentices were not allowed to work more than twelve hours a day.

1833. Factories Act prohibited the employment of children under 9, those aged 9–13 were only allowed to work nine hours a day, and factory inspectors were introduced.

1844. Factories Regulations Act limited the working hours of women to twelve a day.

Gradually legislation improved the working conditions in factories and today they are very much better. All factories have to be registered and conform to the required standards.

1. Premises must be kept clean and refuse not be allowed to accumulate.

2. Workrooms must not be overcrowded, with at least 400 cubic feet ($11.3 m^3$) of space allowed for every person.

3. A reasonable temperature should be maintained. When most of the work is done sitting down the room temperature should not be less than 16 °C (60 °F), and at least one thermometer should be supplied to each workroom.

4. Workrooms should be well ventilated. Also if the work done produces fumes, dust or other substances injurious to health these should be extracted by fans.

5. Adequate lighting should be provided in all parts of the factory. If natural light is used the windows should be kept clean.

6. There must be sufficient lavatories which are properly cleaned, ventilated and lit, and separate for each sex.

7. Sufficient washing facilities with running water, soap and clean towels should be provided.

8. Drinking water should be available, supplied either directly from the public water main, or in special containers protected from contamination.

9. Facilities for hanging up clothing not worn during working hours should be provided.

10. Suitable seating must be provided, according to the job.

11. There should be canteens for workers to eat meals on the premises. Their use is compulsory in factories which use poisonous substances such as lead or arsenic, or where silica or asbestos dust is produced.

12. Factories using large amounts of water should provide adequate drainage channels to prevent water from accumulating on floors.

13. Guards must be supplied to dangerous machinery to prevent accidents. Protective clothing must be provided for certain jobs to prevent injury, e.g. goggles to protect eyesight.

14. First aid equipment must be provided and under the charge of a responsible person who may be required to be trained in first-aid.

15. Reasonable fire precautions must be taken, including means of escape in the case of fire.

FACTORY INSPECTORS

Factory Inspectors have the right of entry to any factory.

1. To check that the employers are aware of and observe the laws regarding the health and welfare of factory workers.

2. To inspect all safeguards against dangerous machinery and chemicals.

3. To check that protective clothing and masks are worn where necessary.

4. To examine the registers in which accidents are recorded. They also investigate the reasons for the more serious accidents.

5. To see that the regulations concerning the employment of young people are enforced.

FACTORY DOCTOR

If the factory is large enough the Chief Factory Inspector will approve the appointment of a part-time or full-time doctor who will have special responsibility for the health of the factory workers and the conditions under which they work. He will be particularly concerned with the health of young people under the age of 18, who will be medically examined to ensure that they are fit for the job they do.

OFFICES AND SHOPS

The Offices, Shops and Railway Premises Act 1963 came into being to safeguard the safety, health and welfare of people employed on these premises. All such premises have to be registered and conform to the required standards.

1. Premises, furniture and fittings must be kept clean and refuse not be allowed to accumulate.

2. Rooms must not be overcrowded.

3. A reasonable temperature should be maintained. When most of the work is done sitting down this should not be less than 16 °C (60 °F).

4. Workrooms must be well ventilated.

5. Adequate lighting must be provided, including in stairways and passages.

6. Sufficient lavatories which are properly cleaned, ventilated and lit must be provided.

7. Sufficient washing facilities with running water, soap and clean towels must be provided.

8. Drinking water must be available.

9. There should be facilities for hanging up clothes.

10. Suitable seating must be provided, according to the job.

11. There should be facilities for workers to eat meals on the premises.

12. Floors, passages and stairs must be properly maintained to prevent accidents.

13. Dangerous machinery, e.g. meat-slicers, must conform to the safety regulations.

14. First-aid equipment must be provided.

15. Reasonable fire precautions must be taken including the means of escape in the case of fire.

WATER SUPPLY

Water circulates continuously between the atmosphere and the land and sea. It falls as rain and may pass through soil, rock and the bodies of plants and animals before evaporating into the air again. The continual re-cycling of water means that it is constantly being polluted and purified.

The average consumption of water is about 250 litres (approx. 50 gallons) per person per day and it is used for drinking, cooking, washing, flushing lavatories, factory processing, fire-fighting, etc. The various Water Authorities throughout the country are responsible for supplying water to their own areas. Each household pays a Water Rate calculated on the rateable value of the house. The water supply must be free from germs and harmful chemicals, and it should also have no smell, taste or colour and be free from suspended matter. But it is desirable that it should contain small quantities of certain chemicals.

Water can be obtained from various sources:

Springs   —Springs at the foot of the South Downs supply Portsmouth.

Rivers   —River Thames supplies Oxford and London.
River Exe supplies Exeter.

Wells   —Peterborough (artesian wells in Lincolnshire limestone).
Brighton (deep wells in South Downs chalk).

Lakes   —Lake Thirlmere in the Lake District supplies Manchester.
Lake Elan in Wales supplies Birmingham.

Reservoirs—Leighton in the Ure Valley supplies Leeds.
Grafham Water, which is filled by the River Ouse, supplies Luton.

Sea Water —Jersey is supplied with distilled sea water.
At Ipswich experiments are taking place on the purification of brackish water for domestic use using techniques other than distillation.

PURIFICATION

Water obtained from upland streams is likely to be safe to drink as it will have had little chance to be contaminated and the ultra-violet rays in sunlight destroy germs. Water from very deep wells is also likely to be safe to drink as germs will have been filtered out as it percolates through the soil and rock to collect underground. In this case care must be taken that the water has not acquired any unacceptable chemicals from the rock which need to be removed before use, e.g. too much lime or iron.

Most water is obtained from rivers, lakes, reservoirs and shallow wells and needs to be purified in Water Treatment Plants to make it safe for use. The treatment the water receives depends primarily on its source although the state of the water from the same source may vary at different times of the year.

WATER TREATMENT PLANT

The diagram (Fig. 16.1) illustrates the basic principles of water treatment. The details vary considerably according to the source and state of water.

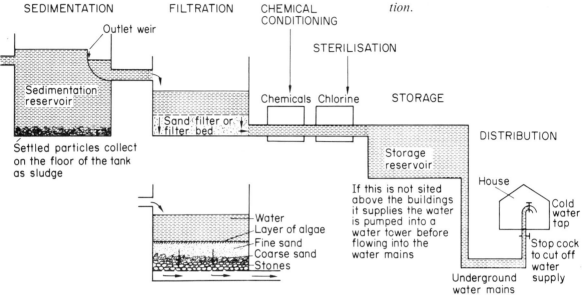

Fig. 16.1 Diagram to illustrate the principles of water treatment and distribution.

SECTION THROUGH A FILTER BED (→ arrows indicate direction of water flow)

Sedimentation

This happens automatically in water stored in lakes and reservoirs but water taken from a river needs to be stored for a while in **sedimentation tanks.** Here chemicals such as aluminium sulphate are added to assist the sedimentation process by flocculation, i.e. the chemicals form 'floc' (like snowflakes) to which the heavier and larger particles adhere. The flow of water through the tank must be at a sufficiently slow rate to allow time for the majority of the particles to settle before the water leaves the tank.

Filtration

The water must next be filtered to remove the finer particles which have remained in suspension. This is generally done in a **sand filter** which is a concrete tank containing a bed of graded sand and stones, usually about one metre deep, under which is a system of collecting pipes. Water from the sedimentation tanks flows onto the sand filter through which it percolates before being collected by the under-drainage system, the finer particles of suspended matter being trapped in the sand. After a period of use the filters become clogged and must be cleared by back-washing with clear water, usually with the assistance of jets of air.

The older method of filtration involves the use of **filter beds** but this is a slow process and the filter beds occupy large areas of land. The filter beds are concrete tanks containing sand and stones graded so that the finest sand is at the top and the largest stones at the bottom. Algae and other micro-organisms live and grow in the upper part of the sand forming a gelatinous layer across the top of the filter bed. Water is pumped onto the bed and gradually filters through to the bottom. Any particles it contained will have been caught up in the gelatinous layer. When this layer becomes too thick it slows up the rate of filtration and has to be removed.

Chemical Conditioning

In certain areas where the water is deficient in desirable chemicals such as lime or fluorine, these may be added at this stage.

Sterilisation

Chlorine is injected into the water flow to kill any harmful organisms which may have escaped the previous processes.

Storage

The purified water is stored until required in enclosed reservoirs to prevent contamination. The reservoirs are sunk in the ground to keep the water cool.

Distribution

It is more economical to site the storage reservoir sufficiently high above the buildings it supplies, but if this is impossible, then the water is pumped up into a water tower before distribution. The water in the tower is thus at a higher level and this causes water to flow along the water mains and through pipes into houses and other buildings, and when the cold water tap is turned on the water will be forced out. Water for the hot water system must be stored in a tank, often in the airing cupboard. Another tank, often in the loft, is needed to supply water for lavatories.

CLEANLINESS OF THE WATER SUPPLY

1. Rules apply to catchment areas, i.e. the areas from which water is obtained, to prevent either people or animals from polluting the land and hence the water.

2. People employed at the Water Works must observe strict rules of hygiene. They are also tested to check that they are not carriers of typhoid.

3. Samples of purified water are regularly tested to check that it is free from harmful germs and chemicals.

4. Care has to be taken when water mains are fractured, to prevent the water becoming contaminated by sewage.

Water borne diseases

Typhoid, paratyphoid, cholera, dysentery, gastroenteritis.

Chemicals in Water

(a) Magnesium and calcium compounds cause 'hardness' which means it is difficult to get soap to lather.
(b) Lead causes poisoning.
(c) Excess nitrate, due to heavy use of fertilisers, causes a blood disorder in babies.

177

(d) Chlorine is added to sterilise water.

(e) Iodine prevents goitre.

(f) Fluorine prevents tooth decay. When this is lacking in the water supply some Local Authorities add fluoride.

(g) Spa water contains various substances which can have a medicinal value.

SEWAGE

Sewage is more than 99% water and comprises all the liquid waste removed from lavatories, baths, wash-basins and kitchen sinks, together with industrial and agricultural waste. It is dangerous to health as it provides a breeding ground for the germs it contains so it is usually removed in sewers to Sewage Works where it is treated to make it harmless.

Sewerage—refers to the system of pipes (sewers).

Sewage—what flows along them.

SEWAGE TREATMENT WORKS

The diagram (Fig. 16.2) illustrates the basic principles of sewage treatment.

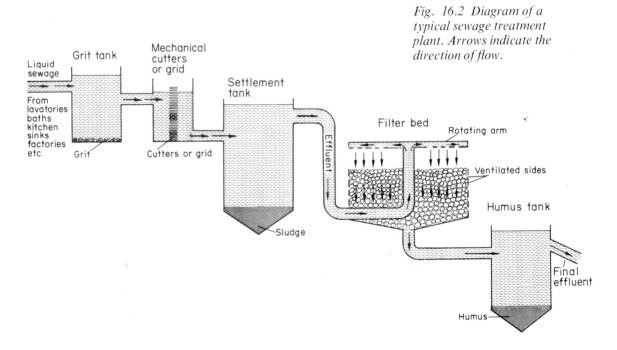

Fig. 16.2 Diagram of a typical sewage treatment plant. Arrows indicate the direction of flow.

Grit Tanks

Heavy inorganic particles such as grit are allowed to sink to the bottom in the grit tanks, and can then be removed.

Mechanical Cutters (or a Grid)

The cutters revolve to shred up paper, rags, etc., into small pieces, or if a grid is used, it acts as a sieve to collect them.

Settlement Tanks

The contents are left undisturbed to allow the heavier particles of organic matter to fall to the bottom where they collect as sludge. The finer particles remain in suspension in the liquid which is called **effluent.**

Filter Bed

The circular filter bed has ventilated sides so that air can enter and it is filled with clinker which allows the air to circulate and effluent to filter through. The effluent is distributed evenly over the surface by the rotating arm above the filter bed. Aerobic bacteria, algae, and other micro-organisms surround the pieces of clinker and use the organic matter in the effluent as food, thus destroying it. Worms and insects move around feeding on the micro-organisms and this prevents the filter bed from becoming choked.

Humus Tank

The effluent leaving the filter bed contains humus, that is, dead plant and animal remains and excreta from the organisms which live amongst the clinker. In the humus tank the contents are left undisturbed to allow the humus to settle at the bottom. The **final effluent** is discharged into a river where any remaining organic matter will be oxidised.

Sludge and Humus

These are drawn off from the tanks as a thick black liquid. Anaerobic bacteria feed on this and in doing so give off foul-smelling gases. To prevent bacterial activity and smells it is necessary to dry the sludge and humus before disposing of it.

METHODS OF SEWAGE DISPOSAL

1. *Water Carriage System* This is the modern method and makes use of water to carry the waste matter through sewers to be disposed of either in a sewage treatment works, or by letting it flow into the sea.

2. *Septic Tank* If a house has a water supply but no sewer, a large tank sunk in the garden can be constructed to receive the sewage. It must be at least 15 metres (50 feet) from a dwelling, have a tightly fitting lid and a ventilation pipe. Anaerobic bacteria will decompose the sewage but when the tank becomes full it must be emptied. This is done by a specially designed vehicle which sucks out the sewage and carries it away for disposal.

Disposal of Excreta Only

1. *Chemical Closet* This is suitable for aeroplanes, boats and caravans, and makes use of chemicals to decompose and deodorise excreta which can be emptied into a cesspit when convenient.

2. *Earth Closet* Earth is used to cover the excreta allowing it to decompose. This is not satisfactory if an earth closet is close to a dwelling as it alters the nature of the soil which surrounds it and renders it unfit for use.

179

HOUSEHOLD REFUSE

This is all the unwanted matter that is put into dustbins or paper sacks and includes waste from eating and cooking, bottles, tins, ash, plastic and paper. The Local Authority arranges for refuse to be collected regularly and it then has the problem of disposing of it.

1. It can be tipped into holes or onto the ground, squashed, and then covered with soil. The organic matter (food, paper, etc.) will rot away and after a while the ground can be used. But the refuse pollutes the ground water, encourages rats and flies, and may smell.

2. If the quantities are large enough, it pays the Local Authority to sort it first to salvage paper, rags, bones, bottles and metal which can then be sold to factories. That which remains will then be either:

    (a) Tipped and covered with soil, or

    (b) Incinerated. Although burning reduces the refuse to a small amount of ash it also produces fumes, some of which may pollute the atmosphere.

POLLUTION

Pollution is a problem in all industrial countries because our modern way of life produces vast quantities of waste matter. This waste causes pollution when it affects the health of the individual, or affects animals and plants, which in turn adversely alters the balance of nature. It is, therefore, important to investigate pollution and to devise means of preventing it or remedying the effects.

POLLUTION OF THE AIR

This is mainly the result of the burning of wood, coal, oil, etc. Smoke, dust and ash are produced as well as invisible gases such as carbon monoxide, sulphur dioxide and other damaging substances. The main sources of air pollution are:

1. Chimney smoke.
2. Exhaust from the motor car.
3. Factory dusts.
4. Cigarette smoke.

Chronic bronchitis is more likely to develop in people who breathe polluted air.

POLLUTION OF RIVERS

This is caused by:

1. Seepage of impurities from refuse tips into the sources of river water.

2. Nitrates used by farmers to fertilise soil being washed into rivers and lakes causing overgrowth of weed and algae.

3. Sewage effluent discharged into rivers lowers the oxygen content of the water which may adversely affect plants and animals living in the river.

4. Industrial effluent containing chemicals and detergents from factories and homes.

Polluted river water will alter the balance of nature and kill fish etc. It is much more difficult and expensive to purify for use in the water supply.

POLLUTION OF THE SEA

This is caused by:

1. Dumping poisonous chemicals in the sea which may be absorbed by fish which are eaten by man. Some poisonous chemicals are harmless in small doses but if they accumulate in the body they cause ill-health.

2. The discharge of raw sewage into the sea. This is unsightly, spoils beaches and bathing, and may be harmful to health.

3. Oil discharged from ships. This floats on the surface and if it is washed onto the beaches it makes them unusable, thus depriving people of a recreational facility.

Oil also creates a particular hazard to sea-birds, notably the swimming and diving species, such as auks, razorbills, guillemots and puffins. The oil penetrates the feathers which lose their waterproofing qualities and the bird becomes waterlogged and its body is subjected to considerable cooling by the water. More often than not the bird is unable to search for food and thus starves to death.

POLLUTION OF THE LAND

This is caused by:

1. Refuse tipping which results in large areas of derelict land. However, if the tipping of refuse is properly controlled the land can be reclaimed and put to good use.

2. Fertilisers used to encourage the growth of crops. If they are unwisely used this may lead to the deterioration of the soil.

3. Pesticides. These are substances poisonous to pests. They must be used with care, especially if they are also poisonous to man, e.g. D.D.T. and Paraquat, or if they alter the balance of nature, e.g. by killing insects necessary for pollination.

NOISE POLLUTION

Traffic, aircraft, industry and radios all produce a considerable amount of noise. The effect of noise on different people varies considerably but generally it makes them less efficient and can be the cause of ill-health.

WORLD HEALTH ORGANISATION (WHO)

Most of the countries in the world belong to this organisation which was set up by the United Nations Organisation after the Second World War. Its purpose is to direct and co-ordinate all matters relating to international health and it deals with problems which need to be tackled on a world-wide scale. Its aim is to bring 'all peoples to the highest possible level of health'.

The headquarters of WHO are in Geneva with regional offices in:

Brazzaville for Africa
Washington for North and South America
New Delhi for South-East Asia
Copenhagen for Europe
Alexandria for the Eastern Mediterranean Region
Manila for the Western Pacific Region.

WHO is financed by annual contributions from the member countries, and by voluntary contributions from them for special purposes. The working languages are English and French, although Russian and Spanish are also widely used. WHO employs doctors, nurses, engineers, administrators, scientists, statisticians, interpreters, translators, secretaries and others.

Some Aspects of the Work of WHO

1. When a country has a health problem its government can ask for assistance and if possible help will be given, usually in the form of specialist advice.

2. It obtains information about infectious diseases to prevent the spread of epidemics.

3. It has launched a major campaign to eradicate malaria.

4. It is concerned with the control of other infectious diseases such as tuberculosis, venereal diseases, rabies and brucellosis.

5. It plays an important role in health education because much disease can be prevented by correct nutrition, cleanliness, vaccination, etc.

6. WHO is also dealing with the problems of drug addiction, alcoholism, epilepsy and mental health.

INDEX

Bold type signifies the main reference.